Manfred Rupalla

A brief insight into the world of device fuses

Significance and Development of Fuses in Electrical Devices

Anchor Academic
Publishing

Rupalla, Manfred: A brief insight into the world of device fuses: Significance and Development of Fuses in Electrical Devices, Hamburg, Anchor Academic Publishing 2015

Buch-ISBN: 978-3-95489-470-3
PDF-eBook-ISBN: 978-3-95489-970-8
Druck/Herstellung: Anchor Academic Publishing, Hamburg, 2015

Cover design by Anna Klöhn
Cover picture by Kai Krueger - Fotolia.com

Translated by Elschukom GmbH, Veilsdorf 2015 - www.elschukom.com

Bibliografische Information der Deutschen Nationalbibliothek:
Die Deutsche Nationalbibliothek verzeichnet diese Publikation in der Deutschen Nationalbibliografie; detaillierte bibliografische Daten sind im Internet über http://dnb.d-nb.de abrufbar.

Bibliographical Information of the German National Library:
The German National Library lists this publication in the German National Bibliography. Detailed bibliographic data can be found at: http://dnb.d-nb.de

All rights reserved. This publication may not be reproduced, stored in a retrieval system or transmitted, in any form or by any means, electronic, mechanical, photocopying, recording or otherwise, without the prior permission of the publishers.

Das Werk einschließlich aller seiner Teile ist urheberrechtlich geschützt. Jede Verwertung außerhalb der Grenzen des Urheberrechtsgesetzes ist ohne Zustimmung des Verlages unzulässig und strafbar. Dies gilt insbesondere für Vervielfältigungen, Übersetzungen, Mikroverfilmungen und die Einspeicherung und Bearbeitung in elektronischen Systemen.

Die Wiedergabe von Gebrauchsnamen, Handelsnamen, Warenbezeichnungen usw. in diesem Werk berechtigt auch ohne besondere Kennzeichnung nicht zu der Annahme, dass solche Namen im Sinne der Warenzeichen- und Markenschutz-Gesetzgebung als frei zu betrachten wären und daher von jedermann benutzt werden dürften.

Die Informationen in diesem Werk wurden mit Sorgfalt erarbeitet. Dennoch können Fehler nicht vollständig ausgeschlossen werden und die Diplomica Verlag GmbH, die Autoren oder Übersetzer übernehmen keine juristische Verantwortung oder irgendeine Haftung für evtl. verbliebene fehlerhafte Angaben und deren Folgen.

Alle Rechte vorbehalten

© Anchor Academic Publishing, Imprint der Diplomica Verlag GmbH
Hermannstal 119k, 22119 Hamburg
http://www.diplomica-verlag.de, Hamburg 2015
Printed in Germany

Table of content

Introduction 7

1. ... until 1860: The research into static electricity 10
2. 1860 to 1900: The golden years of the scientists and pioneers 12
3. 1901 to 1920: Systematization of theoretical principles 21
4. 1921 to 1945: Attempt to establish the fuse as an independent component part 23
   - 4.1. The first specialist companies for safety fuses 24
   - 4.2. The company Wickmann 26
   - 4.3. Limiting current or nominal current? 37
5. 1946 to 1970: Standardization and serial production 40
6. 1971 to 2006: The challenge: Microfuses for "printed" switches 42
7. The basic stages 1981 to 1987 47
8. New approaches in the research until 2012 50
   - 8.1. "The conductor track as a safety fuse: Dangerous bricolage or a matter of necessary, individual safety?" 50
   - 8.2. New wire materials (Company Elschukom) 56
   - 8.3. New theoretic approaches (Company Elschukom) 57
9. List of references and literature 61
10. Chronological overview on the history of the safety fuse 63
11. Work places through the ages of time 67
    - The 1930s 67
    - The 1950s until the 1970s 72
    - The 2000s 75

Introduction

When used in reference to the field of electrical engineering/electronics in general, the term "fuse" describes an trace- or device-related safety measure to protect the operator and environment or component parts from the consequences of residual currents beyond control, by way of shutting down an electrical device before a fire starts and people are exposed to the risk of getting injured due to overheating of pipes and/or component parts.

Over the course of time and its diverse breakthrough moments in history, people have applied various methods. Yet, the basis and at the same time most often component applied has been and still is the safety fuse. In the event of occurring residual current (current, whose occurrence triggers an automatic shutdown, see below), the circuit is cut by the fuse link, which induces the fusing of the fuse element (usually by means of a wire or a strap made of metal like copper or silver) and, in so doing, provides for a "shutdown" of erroneous power source. The fuse link needs to be replaced. At first sight, this may appear to be a disadvantage, which, however, turns out quite useful considering all the aspects related to the safety of both the components and the personnel, as this error won´t occur again when the plant is switched on.

Today, the nearly simultaneously developed "safety switches" or more commonly known as „automatic fuses" are most often used in the field of the plumbing trade. The installation of electrical equipment in living spaces is usually protected by means of automatic fuses, which is a fact that is probably known by most of the readers. Yet, the power supply of apartments to the building connection line is protected using safety fuses.

Residual current

In the field of engineering "residual current" describes a form of current that occurs due to an error in the device, the installation or the supply grid. The term is probably common to the majority of men. For instance, modern installations are equipped with a residual current protective device, which is fitted in the sub-distribution system ("fuse box"). In the event of a short-circuit, the amperage of residual currents may reach up to several hundreds of amps. On the other hand, they may only slightly exceed the normal current rate, e.g. instead of 0.5 A, which would be the case with a premium television set. Provided that this device wouldn´t be equipped with a safety fuse, even the slightly increased current of said premium television might set the device on fire. It´s simple as that: The melting point of the copper in a trace is at 1083 °C.

The importance of fuses and the reason for its development can still be comprehended by way of documented "accidents" resulting from current-related errors.

"According to the fire service, most of the fires are the result of errors in electrical plants and devices or careless handling of electrical household devices. Yet, it has recently been shown that fires may as well be caused by new devices, i.e. such ones equipped with electronic systems like e.g. it is the case with charging sets for cellphones or energy-saving bulbs, PCs etc. (…).. Only too often, power lines become overloaded and, thus too hot. An overload of the electrical power line may also occur, because there are too many devices connected to one plug via several plug connectors."[1]

Most all of these devices are equipped with fuses. It´s likely that the number of house coals would be – other than cited in this paper – which amounts up to 2000 per each year (which, in fact only applies to the city Düsseldorf) – far higher, if the electrical devices were not equipped with a safety fuse, however, the incidents of fire would definitely be lower, if those safety fuses were applied in an appropriate way and with the required expertise.

The two images below depict the consequences that might result from missing or inappropriately fitted fuses by showing an overloaded electric control used for blind motors whereas the other picture shows the experimental laboratory of the Wickmann plants:

---

[1] Federal capital Düsseldorf (Hg.): Risk house fire" – Recognizing hazards, Preventing fires, Düsseldorf 2006 (http://www.duesseldorf.de/feuerwehr/pdf/alle/risiko_wohnungsbrand.pdf <8.2.2012>), P. 4.

Consequences of a "residual current" in a device (Image: M. Rupalla/Wickmann-Laboratory)

Already in 1906 the engineer Georg Isidor Meyer came to the realization of the following: *"The safety fuse can be considered as one of the most sophisticated inventions in the field of electrical engineering. It provides a remedy by triggering the erroneous effect in an enhanced degree. Actually, the reaction is always the same, however compared to the parts to be protected, the effect is considerably increased, thus it can be regarded as an ideal protective device."*[2]

---

[2] Meyer, Georg Isidor: " On the concept of the safety fuse", Munich/Berlin 1906, cited as: Meyer, safety fuse.

1. ... until 1860: The research into static electricity

The history of safety fuses dates back to the 19th century and was characterized by tragic "accidents" that reflected the hazardousness and risks related to the largely unknown field of electricity. Yet, people have known about the existence of static electricity since ancient times, already. It was around 600 BC when the Greek savant Thalas von Milet discovered the static electricity of amber. The attractive force of the charged amber was called "electron" (derived from the Greek term "elkein" = attracting). For whatever reasons, that was it, already. Apart from the "electrostatic generator", which was developed by William Gilbert (1544-1603), who was also the personal physician of the British Queen Elizabeth I in the 16th century, it took until the 18th century until the systematic investigation of the phenomenon commenced by the Dutch physicist Andreas Cunaeus (1712–1788), who invented the "Leydener Flasche", thus one of the first condensers in 1746. This "Flasche" (bottle) was used to collect electric charge, concentrate it afterwards and triggered a heavy electric shock to everyone who got near that point. This process was often compared to the lightning during thunderstorms. After these findings became public, many scientists started to get involved in electric discharges with a continuously growing quantity of the energy that was used for those investigations.

Already in 1774, the scientist Edward Narine used wires, which he aligned with the energy charge in such a way that they were fusing when the energy charge was too high, which resulted in a disruption of the discharge. It is owing to this "alignment" of the wire that laid the foundation for the safety fuse.

This is, what Professor Georg Richmann (1711-1753) should have taken into account. On August 6th, 1753 he was killed while trying to catch "lightnings" using his invented "weather conductor", i.e. a lightning conductor in Petersburg. The investigations regarding the cause of his death showed that the discharging lightning rod struck into his head, went through his body and out of his left foot. Richmann was burned from the inside out. It turned out, that the reason for this was a malfunction in the design of his weather conductor.

Scientists all over the world were shocked of this incident and started to figure out how to avoid such electricity-related accidents in the future. At least since then, the driving force in the development of continuous improvements in the field of safety fuses is the prevention from accidents that might pose a risk to the plants, devices, and equipment, and, above all, to the lives of the staff.

The duration of electrostatic discharges is relatively short, yet, they may be quite powerful and therefore fatal, as the "accident" of Georg Richmann has shown. Besides the analyses conducted with respect to electrostatics, scientists tried to find solutions to provide electricity, which would be available at any:[3]

- 1600 Gilbert (GB): Start of the first electrochemical investigations
- 1789 Galvani (Italy): Discovery of electricity in experiments conducted on animals/frogs
- 1800 Volta (Italy): Discovery of the voltaic cell/"Voltasche Säule" (voltaic pile)
- 1802 Cruickshank (GB): First electric battery in serial production
- 1802 Ritter (Germany): First accumulator „Rittersche Säule" (Ritterian pile)
- 1820 Ampère (France): Electricity using magnetic fields
- 1833 Faraday (GB): Publication of the law by Faraday
- 1836 Daniell (GB): Discovery of the "Daniell-Element"/Galvanic cell
- 1859 Planté (France): Invention of the lead/acid battery

---

[3] GRS Foundation The world of batteries. Function, systems, disposal, Hamburg o.J. (http://www.grs-batterien.de/fileadmin/user_upload/Download/Wissenswertes/ Infomaterial_2010/-GRS_welt_der_batterien.pdf <8.2.2012>), P. 4.

2. 1860 to 1900:
The golden years of the scientists and pioneers

Given those facts, the opportunity and thus the basic pillars for a functional and appropriate use of electricity were already existent before the midst of the 19th century thanks to the galvanic cell and the electromagnetically generated current (current generator). Yet, the opportunity to provide for a permanent supply of electricity entailed the risk of line overloads and short circuits.

Actually, the general use of such power sources was reserved to the field of electric machines like those used for illumination and telegraphy.

All over the world, the electrical engineering was explored, filled with inventions, and put to small and large experiments all the way through to the smallest sphere. Those days, the world was literally overrun by researchers and technicians. Going more into details about these persons would go beyond the constraints of this work.

The chart attached to this work provides a brief (but not complete overview) of the most important names and years

The list contains also the names of German developers and scientists, as they made a considerable contribution to the widespread use of electricity, actually. Examples are the electric engineer Zènobe Thèophile Gramme (1826-1901) – inventor of the direct current generator, i.e. the "Gramme-Maschine" (Gramme machine), which was introduced to the market in 1871 – and the civil engineer Oskar von Miller (1855-1934).

Both set milestones in the field of heavy current engineering. Miller was part of the directive board of the company AEG and was a highly regarded counsellor in the planning processes of electric utilities. Besides that, the exhibitions and fairs whose realization was of vital significance to facilitate the common use of electricity and which took place in 1882 in Munich and 1891 in Frankfurt were organized by him.[4]

Examples for the public use include with the illumination of Berlin's downtown in 1881 and the use of electric motors (navigation on the German Königssee in 1909) and of course the public use of telegraphy.

There are plenty of other examples worth mentioning, however, it can be said that it is the frequent occurrence of "accidents", which direct the focus of this trend towards the indispensability of the development of a fuse system, so that the consequences arising from

---

[4] Info board German Museum of Munich, 2011

overload currents in lines and devices or short circuits can be reduced.

A significant event that emphasizes the importance of fuses is the great fire of Saint Germain in June 1846.

> „In June 1846, because of a fierce thunderstorm over Saint-Germain village, all the wires of Le Vesinet-station were burnt and the apparatus were destroyed. This accident lets us think that we had to protect the operators. We imagine to insert in the electrical circuit a very small and resistant wire, which should burn before the copper wires of the electro-magnets." [5]

The first publication of a full summary on „the use of fuses" ("über den Gebrauch von Sicherungen") dates back to 1874 and was written by Sir David Salomons (1851-1925). Around that time, the development of safety fuses as a means to protect lines and devices.

In 1864, the engineer Sir William Henry Preece (1834-1913) who worked at the "British General Post Office" started to elaborate the issues concerning the maximum capacity of parallel lines. Even today, his findings with respect to the physical laws of the maximum capacity of infinitely long wires are still applied in the development of safety fuses. Although safety fuses were not the focal point of his research work, he was probably one of the first scientists who examined the theoretical principles and requirements related to the processes during the charge and discharge of wires. In collaboration with him, I.M. Onderdonk developed a formula to calculate the shut-down intervals of copper wires. The calculation of these intervals was, however, limited to a rather small amount of time so as to provide for sufficient accuracy (10 s > t > 0.01 s), since the thermal output caused by radiation, lines, and convection was not taken into account.

Later on, this formula provided the basis for the technical specification of the melting integral of a fuse, also known as the I^2t-Wert, which corresponds to the energy required for the fusion.

---

[5] Cockburn, A.C.: On safety fuses for electric light circuits and on the behavior of the various metals usually employed in their construction. J. Soc. Teleg. Eng. 16(1887)5, S. 650-665.

I^2t-value and time-current characteristic

Every time electricity's flowing through a fuse, which brings about the melting of the fuse element, the interval starting with the inflow of the current through to its disruption can be measured and depicted by means of a diagram. The connection line of the measuring points are referred to as "Time-current-characteristic"(I = Current and t = Time).

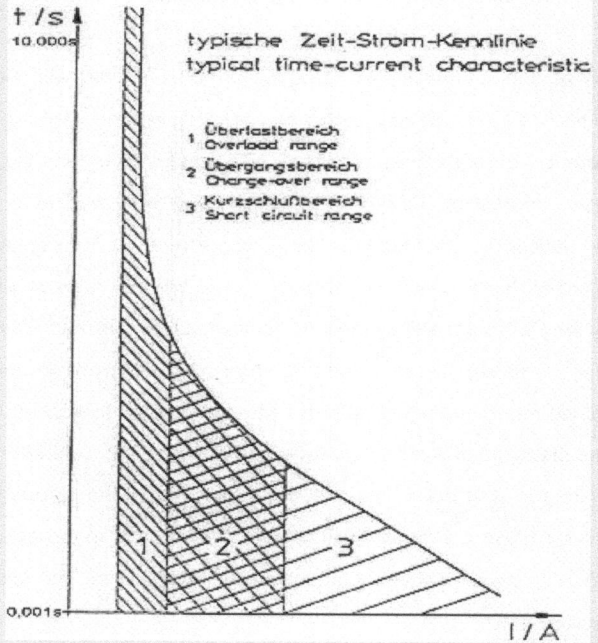

Source: "Wickmann Geräteschutz-Information") 1994

The diagram shows that the „shut-down time" decreases in accordance with the increasing inflow of current. (starting at more than 1.000s in the overload range up to ≤ 0,01s in the short-circuit range). The shut-down time of a fuse in the overload range (1) and the transition section (2) depends as well on the safety fuse's capability of conducting the heat to the connections and surroundings (see explanations on nominal and limiting current). It is only the short-circuit range (3), where the shut-down time is limited to such a short amount of time, that it leads to an immediate disruption of heat conduction. The energy required for the melting process is defined using the I^2t-value. The I^2t-value is of major significance for the technicians and engineers involved in the development of new T.V.-sets, especially when it comes to the choice of the fuse.

Hence, the establishment of a basis for the research and development efforts, which were required to facilitate the common use of safety fuses in times when people started to draw on the application of electrical plants and devices more and more often. Technicians, engineers, and scientists such as Silvanus Phillips Thompson (1851-1916), A.C. Cockburn, Thomas Alva Edison (1847 – 1931) in the 19th century but also G. I. Meyer, E. Wintergerst, L. Vermij, and H. Johann in the 20th century – to mention a few, (L. Vermij listed the names of over 60 scientists in his works) – drew on the paradigms and principles created by W.H. Preece and I.M. Onderdonk to implement their research- and development projects.

It was not until 1879 that electric circuits came with fitted "predetermined breaking points" thus line sections with a smaller diameter. Considering the growing diversity of application fields, Sir P. Thompson realized the necessity of improvements with respect to the safety fuses that had been used until then. It turned out that it was no longer possible to cover the supply of the devices which were supposed to be protected, as the branching of lines in the devices was growing in complexity, i.e. the increasing number of lines resulted in an insufficient supply so that the current flow rate diminished more and more. Hence it was obviously possible to develop thinner lines, yet the risk of a potential overload could not be ruled out. Thompson developed a safety fuse, which shut down the device at low flow rates already, and which were therefore able to be customized to the individual needs of the respective consumer.

He connected to iron wires with a ball made of tin and lead. The resistance of the iron wires generated the heat required to melt the tin-lead-ball, which resulted in a disruption of the line. The resistance of the iron wires provided the basis to determine both the moment when the melting point of the tin-lead-ball was reached and the kind of residual current that led to it:[6]

---

[6] Image source: Gelet, Jean-Louis: To the Origins of Fuses, 8th International Conference on Electric Fuses and Their Application, Clermont-Ferrand 2007, S. 1-8, cited as: Gelet, Origins; here: P. 4.

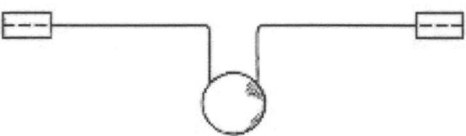

Drawing of the safety fuse developed by Thompson

Not only, this safety fuse design turned out to be an appropriate means to avoid short-circuits but also with respect to the prevention of undesired and hazardous excess currents. Furthermore it opened up the opportunity to align the current intensity (nominal or limiting current – the difference will be explained in detail later on) of safety fuses with the demands of the device or plant to be protected.

That way Thompson set the basic requirements for the safe operation of electrical plants: The selectivity of the applied fuses. This necessity, i.e. the alignment of the nominal or limiting current of a fuse with the lines' cross-section to the conductor or conductors in the device is still considered as the impetus for new developments and improvements of safety fuses.

> ### Selectivity
>
> Safety fuses with all kind of nominal currents are used in all house and apartment buildings. Depending on the number of apartments, the supply line is usually fitted with e.g. 63 A-fuses.
>
> These are used to establish a connection by means of very strong cables (e.g. cables with a diameter of 16 mm^2 per each line). This cable is fused with 32 A-fuses which are installed in the sub-distribution system of the building. All kind of devices (cooker, lamps, dish-washer, ...) or connection points for devices, installed in the apartment (via the plugs) are now protected by means of a 16 A-fuse. Most often, the 16 A-fuse comes in the form of an automatic fuse, which shuts off the power supply in the event of a short circuit in a plug and in doing so, protects the respective line in the apartment (usually those with a diameter of 1.5 or 2.5 mm^2). Provided that the above-mentioned 16 A-fuse wouldn't exist, the upstream 32 A-fuse would probably shut off – or the line installed in the wall would melt, which would eventually lead to a house fire.
>
> The case is similar to the wirings of a plugged-in television. It's very unlikely that the generated flow rate of a defect component part is capable of generating such a high flow rate, which would cause a shut-down through the 16 A-fuse. Yet, it can't be ruled out that the wirings of a television may cause a fire. That's why (almost) all electrical power sources are fitted with additional fuses, whose nominal current is aligned with the failures that may potentially occur in the device. (e.g.. 1 A or 2 A nominal current).
>
> The selectivity principle can be described as follows: Large cable diameter = High nominal current rate of the fuse; small cable or line diameter = low nominal current rate of the fuse.
>
> The 63 A-fuse used in the building connection line must be capable of bearing the total current flow of all connected devices. Given this, it is rather unlikely that the occurrence of a short in a television might cause a noticeable heating in them.

The development of Prof. Thompson was improved and patented by Charles Vernon Boys (1855-1914) and

H.H. Cunyngham in 1883. Cunyngham connected two flat conductors with spring elements using Sn/Pb-solder.[7]

---

[7] Gelet, Origins, S. 5.

Drawing of the safety fuse patented by von Boys and Cunyngham

Like it was the case with Thompson, the tin-lead solder was melted by heating the supply lines, which resulted in a disconnection of them triggered by the spring force. The advantage of the spring force effect was reflected in the operability of the fuse irrespective of its position. Any "dripping"-process with respect to the liquefied solder by means of gravity wasn't necessary any longer.

The wide fields of applications of safety fuses in illumination systems inspired Thomas Alva Edison (1847-1931) and Joseph Wilson Swan (1828-1914), owner of the company "The Edison & Swan United Electric Light Co", to improve Cunyngham's concept of the safety fuse by applying plain wires or belts made of the low-fusing material tin. Later on, in 1881 that is, Edison filed a patent application that contained the term "safety-guard" for the very first time. Indeed, this term is still used today.

The realization of another significant development was achieved by W.M. Mordey. The safety fuses back in those times were mounted in wooden boxes, or as it was patented by T.A. Edison covered with glass. The same goes for the "Bates Fuse", which was fitted with a fuse conductor that was guided through a tube (usually made of ceramics) and whose ends were open.[8]

Not only T.A. Edison but also Mordey found out in 1890 that the molten conductor material of the fuse in a device or switch (these days referred to as "hardware") might be interfering and hazardous. Yet, what was even more important was the control or removal of the arc (also known as "flashover"), which occurs during the melting process of the fuse conductor Due to the increasing diversity of devices, which were connected to a current source and the growing length of supply cables, it became necessary to provide for higher voltages that facilitated the generation of arcs. On the other hand, an uncontrolled shut-down arc switch came along with a higher risk for the environment than an overloaded supply cable, which shut itself off uncontrollably. Mordey led the fuse conductor through a glass tube, which was

---

[8] Andrews, Leonard: Electricity Control. A Treatise on Electronic Switchgear and Systems of Electric Transmission, o.O. 1904.

closed by means of the metallic end caps. The glass tube was filled with sand, chalk, asbestos, or other inflammable materials. That way, i.e. by means of cooling, it was possible to withdraw energy from the shut-down arc and hence to eliminate it in time or to suppress it entirely. Sand with a controlled grain size showed the best results.

Safety fuse by Mordey[9]

Apart from that, Mordey combined several copper fuse elements with the tin foils used by Edison. It´s not known, whether it was intended to establish the nowadays common use of the "M-Effect".[10]

The publication of Mordey's patent in 1890 came along with the introduction of the term "Cartridge fuse". In the time afterwards, many other patents like those from Edison, Mordey and many more were already based on the innovative developments and constructions that were related to the "modern times" as is shown in the figure below:[11]

---

[9] Andrews, Leonard: Electricity Control. A Treatise on Electronic Switchgear and Systems of Electric Transmission, o.O. 1904.
[10] M-Effect = The capability of dissolving copper from liquid tin or lead. Mentioned in the BEAMA-Journal 1939: „A New Fuse Phenomenon" von A.W. Metcalf.
[11] US-Patent Nr. 622,511 – April 1899.

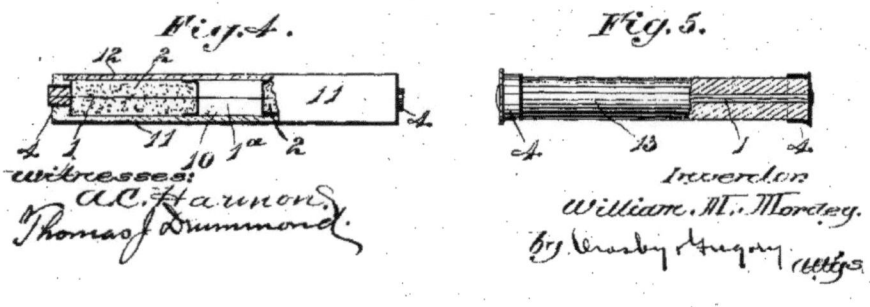

Around 90 later, a safety fuse for the field of telecommunication produced by the company Wickmann looked like as follows:[12]

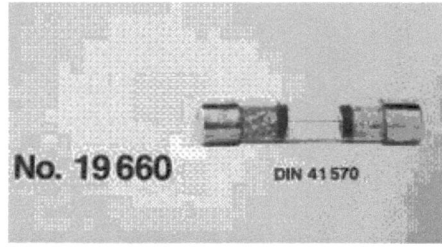

Obviously not much has changed. As is described above, these and other functional principles and constructions of the past can be adopted and thus fit with almost all "developments" of modern safety fuses – .apart from minor exceptions. But more on that later.

In so doing the gap between the "predetermined breaking point" as a rejuvenation of the conductor and the independent component part "safety fuse" was closed. Until the turn of the 19th century the functionality of all commonly used safety constructions and materials was tested and approved.

In order to facilitate the further improvement of the safety fuses, whose functionality had to be aligned with the growing diversity of devices and applications, it was necessary to elaborate the functional principles and the description of the respective physical law with respect to the processes before and after the melting of the fuse conductor.

---

[12] Wickmann-Werke AG: Catalog, Witten 1980 (Copy in possession of the author.).

3. 1901 to 1920:
Systematization of theoretical principles

Until 1900 there were only a few theoretical papers, which aimed at conveying a better understanding of the physical processes during the melting of a safety fuse conductor. Apart from the above-mentioned publications by the engineers W.H. Preece and I.M. Onderdonk back in 1864 and those by Herzogs and Feldmanns "Die Berechnung elektrischer Leitungsnetze in Theorie und Praxis" (The calculation of electrical wiring system in theory and practice) from 1893, the important basic requirements to realize improvements with respect to the safety fuse were missing. Although their publications were quite interesting, the renowned engineers Forbes, Reinisch, Grassot, Skrischinski, Uppenborn and Oelschläger were not able to provide a more revealing insight into the topic. As of 1900, the number of the scientists and engineers who have been dealing with this topic started to push the boundaries; so let's have a look at the developments and trends in Germany. It should be mentioned that all developments taking place over here were tightly connected to the research in other countries.

In 1906 Georg Isidor Meyer published his above-mentioned paper: "Zur Theorie der Abschmelzsicherung".[13] (About the theoretical principal of the safety fuse) and did not only investigate the processes taking place with respect to the heat distribution and heat conduction during the heating of a wire, but also the compatibility of various materials and their combinations by means of several different construction types. He analyzed the behavior of parallel fuse elements, coiled wires and wires with constrictions. He was looking for regularities in different kinds of impacts during the heating of all those designs and materials. Besides numerous formulas for the calculation of fusion characteristics, Meyer discovered constants, which are typical for every kind of material. In doing so, it was possible to calculate the melt flow rate of high flows or very short intervals in a simple way. Even today, this "Meyer-Constant" is applied by the developers of the safety fuse manufacturers to align them with the existing applications or (primarily) standards. Despite all the ground-breaking and innovative discoveries, the author Meyer described in his paper proved the voltage isn't important when it comes to the disconnection of the fuse. Most of the tests were conducted using direct current voltage below 10 V. Not only this was technically incorrect but also it considered merely a part of the applications.

---

[13] Meyer, Safety fuse, P. 1.

Due to the strongly varying levels, the increase of the significance concerning the application of higher operating voltages is rather unlikely, although the "Zed"-fuse (similar to today's fuses used in house wiring systems), which was developed and produced by Siemens for the purpose of generating higher voltage levels back in 1912, was sold more than 250,000 times in the UK.

G.I. Meyer's opinion concerning higher voltage levels:

> "The very peculiar results of research works were published by W. Oelschläger (Elektrotechn. Zeitschr. 1904 -- Journal for Electrical Engineering). He conducted research on the melting process of a fuse in current- and voltage amplifications during short circuits using an oscillograph. (...) The present paper discusses the principles of the safety fuse in general without examining the effects after the melting process, so that the principles applying for the practical operation, thus resistance, self-induction, and capacity with the voltage merely exerting a relatively minor impact (...)." [14]

Meyer continued, amended, and complemented the works of (e.g. Preece and I.M. Onderdonk) in all areas. However, after all it's only his findings in the field of the material characteristics, which are still applied in our development departments and by the manufacturers of safety fuses, although their theories and principles have fallen into oblivion, already. Nevertheless, it is obvious that Meyer's contemplations regarding the significance of voltage were full of mistakes.

---

[14] Meyer, Safety fuses, P. 4.

4. 1921 to 1945:
Attempt to establish the fuse as an independent component part

Due to the rapidly spreading use of electrical plants and devices, the demand for safety fuses increased as well. First, the manufacturers of power distribution systems, lighting equipment, motors, etc. (Siemens, AEG, Meyer et al.) developed and produced the required safety fuses on their own, but at some point they were no longer in the position to provide for the supply of the replacement demand. Other manufacturers were found soon and, most importantly they agreed to acquire the knowledge about safety fuses and to take on the repair and maintenance of switched-off fuses. Since the use of open fuses was widespread in Germany, repair and maintenance works were not only reasonable but also possible. Two of those various construction types are shown in the figure below:[15]

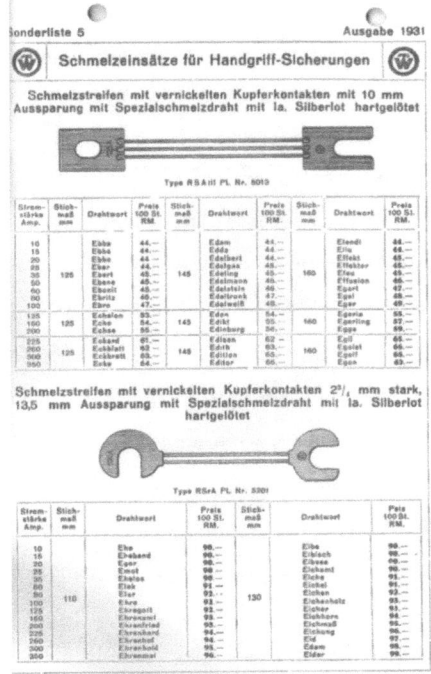

---

[15] Wickmann-Werke AG: Catalog, Witten 1932 (Copy in possession of the author)

Those safety fuses were easy to repair and there was, in fact, a great demand for it. However, it is rather unlikely that the frequently mentioned "laminated fuse" as it was offered in a catalog by the company Pudenz in 1937, was overhauled, too.[16]

As far as is known, this type of fuse was actually manufactured by Pudenz and Wickmann. Unfortunately, there haven´t been found any records like data sheets or specifications on the laminated safety fuse and its various types.

4.1. The first specialist companies for safety fuses

It can be assumed that the company "Pudenz KG", which was established in Wuppertal back in 1909, was one of the first German companies that specialized in safety fuses and others. In the first years of their presence on the market, the focus of their activities was on the repair and maintenance of safety fuses. The growing knowledge about the functionality of safety fuses might have been an incentive for them to take on development tasks, as well. The way of how they developed and established their business is similar to the company Wickmann, which was founded in Dortmund in 1918. While the company Schurter KG in Luzern (Switzerland) entered the market of "safety fuses" through the repair and maintenance of the latter, the companies Bussmann (St. Louis, Missouri, USA, 1914) and Littelfuse (Chicago, USA, 1927) started to implement their development plans for safety fuses (see overview attached to this document).

The use of electrotechnical devices increased at a rapid pace in the USA, Europe, and hence Germany. One of the primary reasons for the high demand of fuses can be ascribed to the developments in the fields of telecommunication and radio industry, as the protection of single devices was attached with increasing importance. The commercial use of this technology got into its stride.

Back in 1903, the "Gesellschaft für drahtlose Telegraphie m. b. H (Association for wireless telecommunications) which was based in Berlin, were the founders of the two German electricity companies Siemens & Halske A.G. and the Allgemeine Elektrizitäts Gesellschaft, Berlin. According to their principles and goals with respect to technology, that new association

---

[16] Copy in possession of the author.

represented the fusion of the systems by Braun-Siemens and Slaby-Arco dar. The new System was called ‚Telefunken'. Count Arco was appointed to be the technical executive and principal engineer of the newly established company. (…)."[17]

Back in 1915 Guglielmo Marconi succeeded in establishing a telephone line to a ship that was about 50 km away. In the early 1920s, several radio stations signed on (1919 Nauen, 1924 Leipzig …). Despite the fact that merely about 20 percent of the German population were able to use electrical power in 1928, the popularity and demand for radio sets was considerably high: *"On the 1st of April 1924, the number of listeners in Germany amounted to 9895, one year later, the number had increased to a whopping 872.600, 1.246.000 at the beginning of the year 1926, and eventually a whopping 2.334.000 of listeners in October 1926."*[18]

The repairable safety fuses applied in the early stages of the century were – at least in Germany – open fuses with voltages[19] varying from 10 A to 600 A. In order to comply with the principle of "selectivity", the installation of safety fuses for single device or antenna systems of houses with voltages varying from less than 10 A down to 0.01 A became necessary. This came along with specific developments and designs, which had to be adopted by the then existing manufacturers of safety fuses manufacturers. This separation in the development of safety fuses is assumed to be the reason, why the fuse was no longer recognized as a component part by the engineers of devices. An example for this is a wiring diagram: Published in the magazine "Funkschau" on 30th of April 1939 and set up for the radio set "Rekordbrecher-Sonderklasse" the safety fuse was not referred to as a component part as incidentals.[20]

In the late 1920s the Witten-based company Wickmann-Werke AG started to develop and manufacture safety fuses. The company Wickmann wasn´t the first one that developed safety fuses, yet it became the most important supplier of fuses in Europe. As is described in the following section, the most significant impulses for innovation in the field of research and development, as well as those related to the standardization originate in the works of Wickmann in Witten, Germany.

---

[17] Telefunken-magazine Issue. 17/1919, P. 10 (http://www.Radiomuseum.org <8.2.2012>).
[18] The first radio set 1924. Publication in the "Freie Presse" from 3.3.1992 (http://www.krumhermersdorf.de/-geschichte/k-g1631.htm <8.2.2012>).
[19] Actually the correct definition of the colloquial term "voltage" is "nominal current". For further details, refer to the explanations: "Nominal current or limiting current".
[20] Funkschau magazine 18 – April 1939, "Der Kontrastheber Rekordbrecher-Sonderklasse", P. 130, Heinz Boucke (Wiring diagram of the radio set "Rekordbrecher-Sonderklasse").

4.2. The company Wickmann

In 1918 the company Wickmann broke into the market of repairs and maintenance services for safety fuses. It didn't take too long, before they gained a foothold in the field of developing and manufacturing safety fuses. Their range of products wasn't limited to all types of safety fuses. It did also include auxiliary materials required for the construction of antennas and lightning arresters, or even voltage gauges, surge diverters, and fuse holders. Both have been developed and manufactured since the 1930s through to the third millennium. Later on, the company added thermal fuses and heavy-duty switches to their product range.

Online researches at the German Patent Office showed that the number of patents registered by Wickmann from their early years back in the 1930s through to the close down of the company amounted to 660. Apart from those, he applied for various other patents in other countries or Euro-patents.

It goes without saying that much importance was attached to the product diversity. Wickmann established itself as the leading supplier and manufacturer of pretty much everything that contributed to the protection of plants and devices. Although Wickmann wasn't the "inventor" of the equipment fuse, he can definitely referred to as the inventor of the unmistakable safety fuses. Already in 1932, Wickmann launched a system of safety fuses, which could be used to provide all types of fuses (with nominal currents from 10 mA to 6 A) with "adjustment fuses". Since the nominal current range consisted of approx. 30 nominal current levels, they were broken down into different lengths.

- o Nominal current voltage 10 mA – 60 mA   Total length 30 mm
- o Nominal current voltage 80 mA – 600 mA   Total length 25 mm and
- o Nominal current voltage 700 mA – 6000 mA   Total length 20 mm"[21]

---

[21] Information by Jost Degen, company archive Schurter, 2008.

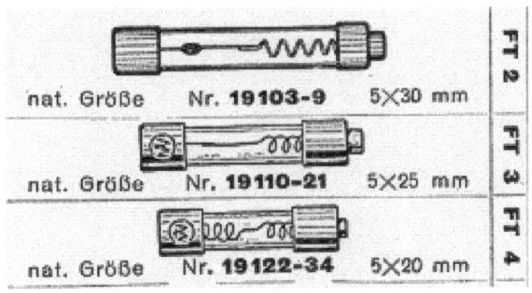

Fuses with adjustment pins[22]

That way, it was possible to make sure that people (especially non-professionals) used the appropriate fuse along with the corresponding nominal current, when it needed to be replaced The improvement and production of those fuse components, such as the surge diverter ("lightning protection") was continued until the 1990s.

The image below shows a "surge diverter" taken from a Wickmann catalog dating back to 1933:[23]

---

[22] Wickmann-Werke AG: Catalog, Witten 1932 (Copy in possession of the author)
[23] Wickmann-Werke AG: Catalog, Witten 1932 (Copy in possession of the author.).

Sonderliste 20   Ausgabe 1933

Wickmann-Störschutz-Ueberspannungsableiter

Sämtliche Armaturen für Antennen mit Störschutzkabel müssen abgeschirmt sein und dürfen nicht mit normalen Überspannungsableitern bezw. Blitzschutzgeräten versehen werden. Vielmehr ist der besonders abgeschirmte

Wickmann-Störschutz-Ueberspannungsableiter

erforderlich, der, wie aus der Abbildung hervorgeht, vollkommen metallisch gekapselt ist.

Das Gehäuse besteht aus einer guten wetterbeständigen Spritzgußlegierung. Die Einführung des Kabels erfolgt mittels Verschraubungen mit gleichzeitiger Zugentlastung. Die konstruktive Ausbildung macht ein Vergießen der Kabeleinführung überflüssig.

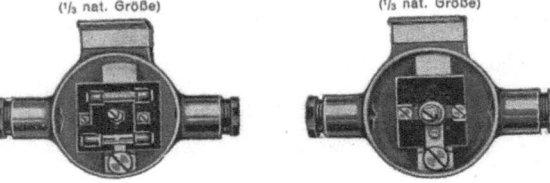

                    Type SP I   Type SP II

Das Gerät Type SP I ist mit einer empfindlichen Feinfunkenstrecke in Form einer Edelgaspatrone, die schon bei ca. 100 Volt anspricht, versehen. Parallel zu dieser Edelgaspatrone ist eine Grobfunkenstrecke angebracht. Der Erdungsanschluß ist gegen Gehäuse bezw. Abschirmung isoliert. Diese Isolierung ist durch eine Schraube metallisch überbrückt. Soll die Abschirmung als Gegengewicht benutzt werden, so ist diese Schraube zu lockern, wodurch eine Funkenstrecke entsteht. Fernerhin ist in die Leitung zum Empfänger noch eine Schmelzsicherung eingebaut, die bei größeren Überspannungen (Blitzeinschläge!, Übertritt von Starkstrom in die Antenne!) sofort zum Abschmelzen kommt und den Weg zum Empfänger sperrt.

Das Gerät Type SP II besitzt einen gemeinsamen Erdungsanschluß für zwei Feinfunkenstrecken, die gemäß den VDE-Vorschriften bei 350 Volt ansprechen. Die Ableitung der Überspannungen von der Antenne zur Erde erfolgt über eine dieser Funkenstrecken. Die andere Feinfunkenstrecke dient zur Ableitung der Überspannungen aus der Abschirmung. Außerdem ist noch eine Grobfunkenstrecke zwischen Abschirmung und Erde vorgesehen, die wie bei Type SP I mittels einer Schraube kurzgeschlossen ist. Wird die Abschirmung als Gegengewicht benutzt, so ist diese Schraube zu lockern.

Beide Ausführungen stellen eine ideale Lösung der Entstörung des Empfanges dar und bieten in weitgehendem Maße einen wirksamen

Schutz für Eigentum und Leben!

Surge diverter dating back to *1933*

Sonderliste 19   Ausgabe 1936

Aufstellung der wichtigsten Empfänger-Fabrikate mit Angabe der Absicherung der einzelnen Geräte. "Wickmann-Feinsicherungs-Packungen"

Fabrikat und Apparate-Type	bei Betriebs-Spannung Volt	erforderliche Feinsicherung Strom-Stärke mA	Größe mm	Fein-sicherungs type	Drahtwort für Fein-sicherungen	Preis 1 Stück ℛℳ	Feinsicherungs-Packungen, enthaltend 2 Feinsicherungen Drahtwort für Packungen	Preis 1 Pack. ℛℳ
A E G (Fortsetzung) AEG 2M¹⁶W, AEG 2¹⁶W, AEG 3³⁵W, AEG Super4⁵⁶W, AEG Groß-Super 4⁷⁶WK, AEG Luxus-Super 6⁹⁸WK	220/240	Lötstr.	42 lg.	19±05	Traut	—.18	Trautpa	—.36
Beteco								
1935								
Beteco-Halle W	110/220	700	5×20	FT4	Tretam	—.30	Tretampa	—.60
Beteco-Extra	220	700	5×20	FT4	Tretam	—.30	Tretampa	—.60
Betecophon W	110/220	700	5×20	FT4	Tretam	—.30	Tretampa	—.60
Beteco-Riese	110/220	1000	5×20	FT4	Tretip	—.30	Tretippa	—.60
Blaupunkt								
1929 NKW1.5, NKWR1.5, NS IV	110/220	260	5×20	FN1	Tambur	—.30	Tamburpa	—.60
1930 W300, LW300, G300, LG300	110/220	400	5×20	FN1	Tantal	—.30	Tantalpa	—.60
NTS5	110/220	0,8 A	5×20	FN1	Taric	—.30	Taricpa	—.60
1931 W400, LW400, G400, LG400, VLG400	110/220	0,8 A	5×20	FN1	Taric	—.30	Taricpa	—.60
1932 LW2000, LG2000, W4000, LW4000, LG4000, LW4004	110/220	0,8 A	5×20	FN1	Taric	—.30	Taricpa	—.60
1933 LW3000, LG3000, Super4LGH, 4LGP	110/220	0,8 A	5×20	FN1	Taric	—.30	Taricpa	—.60
Super4LWH, 4LWP	110/125	1 A	5×20	FN1	Tai	—.30	Taipa	—.60
	150/220	0,6 A	5×20	FN1	Taon	—.30	Taonpa	—.60
1934 Super4G6, Super3G4, Super4W9, Super4W6, Super3W6, Super2W2, Super3W4	110/220	0,8 A	5×20	FN1	Taric	—.30	Taricpa	—.60
	110/125	1 A	5×20	FN1	Tai	—.30	Taipa	—.60
	150/240	0,6 A	5×20	FN1	Taon	—.30	Taonpa	—.60
1935 4W95, 4W65, 4W55	110/130	1,2 A	5×20	FN1	Tabil	—.30	Tabilpa	—.60
	220/240	0,7 A	5×20	FN1	Tageis	—.30	Tageispa	—.60
4GW65, 3G15	110/240	0,8 A	5×20	FN1	Taric	—.30	Taricpa	—.60
3W15	110/130	1 A	5×20	FN1	Tai	—.30	Taipa	—.60
	220/240	0,6 A	5×20	FN1	Taon	—.30	Taonpa	—.60
1936 Super5W86, Super4W76	110/130	1,2 A	5×20	FN1	Tabil	—.30	Tabilpa	—.60
	220/240	0,7 A	5×20	FN1	Tageis	—.30	Tageispa	—.60
Super4W66, Super4W65K, Super3W56	110/130	1 A	5×20	FN1	Tai	—.30	Taipa	—.60
	220/240	0,6 A	5×20	FN1	Taon	—.30	Taonpa	—.60
Super4W76, Super4GW56	110/220	0,8 A	5×20	FN1	Taric	—.30	Taricpa	—.60
Brandt								
1932 W56, W64, LW72, LW80	110/220	260	5×20	FN1	Tambur	—.30	Tamburpa	—.60
1933 Columbus 94 und 100	110/220	260	5×20	FN1	Tambur	—.30	Tamburpa	—.60
Columbus W153 und LW180	110/220	300	5×20	FN1	Tango	—.30	Tangopa	—.60
Columbus W195	110/220	600	5×20	FN1	Tarif	—.30	Tarifpa	—.60

— 17 —

Safety fuses for specific construction types

The increasing electrification of private households brought about a constantly changing sensitivity with respect to the required safety fuses. Especially the radio sets of the new generation required lower and lower nominal current levels with more and more diverse characteristics. Not only the nominal current, but also the shut-off characteristics were aligned with the respective devices.

Those developments included various construction types, which were compatible with different voltages or fuse elements so as to enable the fast (quick) or slow (idle) deactivation of a residual current.[24]

On the one hand, this was a quite reasonable solution, yet it led to the accumulation of „special designs", which has become not even remotely comprehensible.

The "Volksempfänger" introduced at the IFA in Berlin on 18th of August 1933 was supposed to be as reasonable as possible. Hence, the same applied to safety fuses. The only way to realize this was through serial production, however, given the partly very complex construction in relation with customized safety fuses made the realization of it impossible in the first place.

In general it can be said that the technical progress went on rapidly in the 1920s and 1930s. Individual equipment or component parts were designed and manufactured for each plant, device, or machine. Protective devices including fuses were certainly part of it as is shown in the abstracts of the Wickmann catalog:[25]

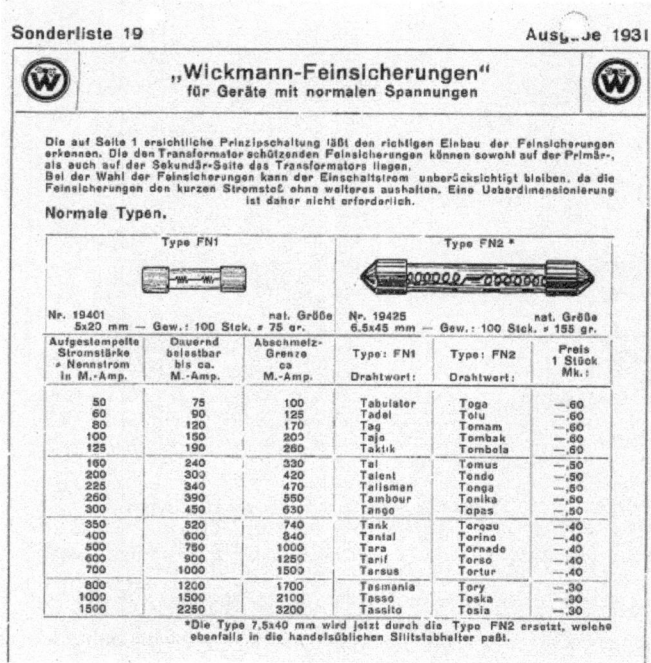

---

[24] Wickmann-Werke AG: Catalog, Witten 1932 (Copy in possession of the author.).

[25] Wickmann-Werke AG: Special item list 19, Witten 1931 (Copy in possession of the author.).

Sonderh..e 19   Ausgabe 19

 Aufstellung über die in den verschiedenen Apparaten verwendeten Feinsicherungen.

In nachstehender Tabelle bringen wir den Wünschen vieler Radio-Händler entsprechend, eine Zusammenstellung der Apparate-Typen der bekanntesten Netzanschlußgeräte, aus der zu ersehen ist, mit welchen Feinsicherungen die einzelnen Apparate abzusichern sind. Die gemachten Angaben sind unverbindlich, Preise etc. siehe Seite 2 dieser Liste.

"Owin" Feinsicherung: Type FN2 = 6,5x45 mm		"Tefag" Apparat-Type		Feinsicherung: Type FN4 = 8x30 mm abzusichern bei:		
	abzusichern bei:			Wechselstrom		Gleichstrom
Apparat Type:	110-127 Volt mit M.Amp.	165-220 Volt mit M.Amp.	Nr.	Bezeichnung	110, 125 bis 160 Volt mit Amp.	220 Volt mit Amp.	220 Volt mit Amp.
E16W	260	160	1350	Tefaphon	0,8	0,4	1,5
E23W	260	160	1358	Kraftverst.	0,8	0,4	—
E11W	260	160	1304	Tefagon 4	0,8	0,4	2x0,4
E22W	260	160	1406	Anod.-Netzg.	0,8	0,4	—
E26W	260	160	1315	Tefagon 34L	0,4	0,2	2x0,4
V25W	260	160	1368	Kraftv. K 1,8	1,2	0,5	2x1,5
V22W	260	160	1351	Truhn T 12	1,2	0,6	2x1,5
V20W	500	260	—	—	110,125 bis150Volt		—
V24W	500	260	1318	Tefakkord	0,4	0,2	2x0,4
V14W	1000	500	1319	Tefakron	0,4	0,2	2x0,4
VL1W	500	260	1320	Tefadyn	0,4	0,2	2x0,4
VL2W	500	260	1353	Tefariston	0,4	0,2	—
G6W	260	160					
G7W	260	160	—	—	110u.125 Volt	150Volt	—
V20G	—	1000					
VL1G	—	1000	1364	Kraftv.E K12	1,2	0,8	0,6	—

Thanks to the German industry standard, (DIN), which has been existing since 1918 in order to provide for the reproducibility and replacement of parts in general, almost everything was standardized. In March 1918, the tapered dowel pin of the machine gun 08/15 was standardized in the first DIN standard – DIN 1. According to the notes and references at hand, the standardization of safety fuses or device protections was generally limited to their designs and dimensions.

Actually, the regulations concerning the "Zur sicheren Erstellung elektrotechnischer Anlagen" (On the appropriate and safe engineering of electrotechnical plants) – VDE0100 by the VDE (Verband Deutscher Elektrotechniker) (Association of German Electrical Engineers), which was founded in 1893 were published already in 1895, yet, those rules merely defined the protection of lines. Due to the diversity of devices, the voltages of the fuses required were much lower usual with lines connected to apartment buildings and houses.

In the 1930s, Wickmann´s chief engineer Oskar Ackermann realized the necessity of further standardization measures. This is, what he described in the conclusion of his paper "Warum eine Gerätesicherung?" (What´s the purpose of a safety guard?) from 1936:

> „Jedes Gerät, das beim Eintritt eines Schadens eine erhöhte Stromaufnahme bedingt, die zum Ansprechen einer Stromkreissicherung nicht genügt, muß mit einer Gerätesicherung versehen sein, um das Gerät und gegebenenfalls dessen Umgebung vor Schaden oder auch gänzlicher Zerstörung zu schützen."[26]

(Every device that causes an increased power demand in the event of a damage and whose power isn´t capable of triggering a circuit protection mechanism must be provided with a safety fuse in order to protect both the device and its environment from damages or, in the worst case, its destruction.)

And furthermore... "Eine Sicherung, die den heutigen Erfordernissen entspricht, muß folgende Bedingungen erfüllen: (A safety fuse that is supposed to comply with today´s demands, must meet the following requirements)

a) The melting curves must be aligned with the device.
b) The fuse must provide the appropriate inertia.
c) The fuse must be highly responsive to shorts.
d) (...)
g) The fuse must be capable of suppressing the generation of an arc when the device is shut off."[27]

The use of customized safety fuses was vital to meet the high requirements (see data sheets above). It was almost at the same time that Ackermann emphasized the importance of the standardization of essential fuse characteristics, e.g. the voltage or the nominal resp. limiting current, as well as the test conditions. In a lecture he held in the "Reichsluftfahrtministerium" back in 1936, he addressed this problem and proposed a solution for it:

> „Die bisherige Entwicklung der Flugzeugsicherungen brachte es mit sich – wie bei jedem neuen Gebiet –, dass jeder Flugzeugkonstrukteur die Nennstromstärken und auch die Abschmelzstromstärken für seine Sicherungen in einem Flugzeug bestimmter Bauart festleg- te. (...) Dieses Verfahren erschwerte die Beschaffung und Lagerhaltung außerordentlich. Im Ernstfalle würden solche Umstände dazu führen, daß eine Einsatzbereitschaft außerordent- lich gefährdet würde."[28]

(The past experiences in the development of safety fuses for planes showed – as it is the case with every new field – that both the nominal current voltages and the melting voltages for the corresponding fuses of type-specific planes were determined by the plane engineer(...) This method made the procurement and the stock-holding very difficult. In the worst case, such circumstances might compromise the operability of the device.

---

[26] Ackermann, Oskar: Warum eine Gerätesicherung? (The significance of safety fuses), Witten 1936.
[27] Ackermann, Oskar: Die Absicherung von elektrischen Geräten (The protection of electrical devices), Witten 1934.
[28] Ackermann, Oskar: Die Absicherung elektrischer Anlagen in Luftfahrtzeugen" (The protection of electrical devices and devices in aerial vehicles) Witten 1936.

Ackermann introduced a "series of standards" for plane safety fuses, which set the basis for nominal currents and test specifications after "extensive consultations and analyses". The text contains no details concerning the "who asked whom for advice", yet it can be assumed that the consultation for "expert advice" took place on site of the company Wickmann.

The above-mentioned series of standards for plane fuses was already subject to various nominal currents and time-current characteristics, which were bound to specific times depending on the various current loads. However, the paper reveals that the standardization sheets 500xxx (time-current characteristics) and FL327xx (dimensions and nominal currents) are based on specifications provided by the company Wickmann itself.

Communication media have been subject to standardization since 1940 – yet, the standardization process was very time-consuming and the focus was on keeping pace with the speed of the development:

> „Die Normung der Nachrichtenmittel hat ihren Niederschlag in der Form der deutschen Einheitsblätter (DIN-E-Blätter) gefunden, die der Deutsche Normenausschuß 1940 neben den eigentlichen Normblättern (DIN-Blätter) zur Abkürzung des Bearbeitungsverfahrens einführte. Diese Einheitsblätter können ohne das bei gewöhnlichen Normblättern übliche, meist sehr zeitraubende Veröffentlichungsverfahren verabschiedet werden. In ihrem Inhalt unterscheiden sie sich nicht von gewöhnlichen Normblättern. Als äußeres Merkmal tragen sie lediglich zwischen „DIN" und der Ordnungszahl das Wort „Einheitsblatt". Im Laufe der Zeit soll, wenn die Entwicklung der Normteile einen gewissen Abschluß erreicht hat, von Einheits- auf Normblätter übergegangen werden."[29]

(Instituted by the German Committee for standardization in 1940, the standardization of communication media was implemented by means of the German standard sheets (DIN-E-Sheets), which, other than the usual specification sheets, allowed for a shortening of the usually time-consuming process. Actually, there's no difference between these and the other, regular standard sheets. The only thing that distinguishes them from the regular standard sheets is the addition of the word "standard sheet" between the word "DIN" and the corresponding ordinal number. The adoption, thus the change from the old standard sheets to those ones described above is supposed to take place after the development of the parts to be standardized has reached a certain point of progress.)

Although the glass-tube- and laminated fuses that have been in use for power outlets since 1925 are defined through DIN 49398 later VDE 0610/4, §54 Schmelzeinsätze für Steckdosen (fuse links for power outlets), it took 18 years, i.e. until April 1943 for the general VDE-standard

---

[29] Cames, Wilhelm: Die Normung in der Funktechnik, in: Funkschau 17 (1944) 3/4, S.17-18.

0820 "„Leitsätze für Gerätesicherungen der Fernmeldetechnik" (Principles for safety fuses in the field of telecommunications) to gain approval and recognition. The previously common complex and constantly changing stock of standards was indirectly described by Wickmann´s chief engineer Hermann Bellen:

> „VDE 0610/4.55 enthält von der außer Kraft gesetzten Vorschrift VDE 0610/1.45 nur noch den §54 betr. Schmelzeinsätze für Steckdosen, jedoch gilt statt DIN 49398 jetzt DIN 41571 Bl. 1 und statt des Prüfverfahrens nach §51 ist VDE 0635 anzuwenden".[30]

(Apart from the suspended provision VDE 0610/1.45, the regulation VDE 0610/4.55 merely contains §54 regarding fuse links for power outlets, yet DIN 49398 has been replaced with DIN 41571 Bl. 1 and the test procedure according to §51 has been replaced with the regulations set out in VDE 0635.)

---

[30] Hermann Bellen, „Die Gerätesicherung", Witten 1955.

Leitsätze für Gerätesicherungen der Fernmeldetechnik

VDE 0820/XII. 43

Mit VDE 0820 U/I. 43 — Nachdruck 1952

Inhalt

I. Gültigkeit
§ 1 Geltungsbeginn
§ 2 Geltungsbereich

II. Begriffe
§ 3 .

III. Baubestimmungen
§ 4 Normen

§ 5 Aufschriften, Nennspannungen, Nennströme
§ 6 Bauvorschriften
§ 7 Kriech- und Luftstrecken

IV. Prüfung und Bewertung
§ 8 Schmelzeinsätze
§ 9 Sicherungshalter

I. Gültigkeit

§ 1
Geltungsbeginn

a) Diese Leitsätze treten am 1. April 1943 in Kraft[1]).
b) Für bestehende Anlagen bleibt die Verwendung bisher üblicher Gerätesicherungen zulässig, wenn ihre Verwendung nicht ausdrücklich durch eine Bestimmung des VDE für unzulässig erklärt ist.

§ 2
Geltungsbereich

a) Diese Leitsätze gelten für alle Gerätesicherungen für die Fernmeldetechnik, d. h. für Fernmeldeanlagen nach VDE 0800, für Fernmeldegeräte nach VDE 0804 und für Rundfunk- und Schallgeräte nach VDE 0860.
b) Sie gelten vorerst nicht für rücklötbare Sicherungen und Spannungssicherungen bzw. -ableiter.

II. Begriffe

§ 3

a) Gerätesicherungen als Stromsicherungen sollen Geräte und Anlagen gegen Ströme unzulässiger Stärke und

---
1) Genehmigt durch den stellvertretenden Vorsitzenden des VDE im Januar 1943. Bekanntmachung: ETZ 1943 S. 113 (Einführung hierzu ETZ 1943 S. 116). — 1. Änderung (gültig ab 1. März 1944) genehmigt durch den Vorsitzenden des VDE im Dezember 1943. Bekanntmachung: ETZ: 1943 S. 567 und 1944 S. 43.

Verband Deutscher Elektrotechniker e. V.

Die vorliegende Fassung ist verbindlich, bis eine Änderung oder Neufassung gemäß einer Bekanntmachung in der ETZ in Kraft tritt.

VDE-Verlag GmbH, Wuppertal und Berlin-Charlottenburg 4

Abstract VDE 0820[31]

---
[31] DKE Deutsche Kommission; Genehmigungs-Nummer 202.013

Not only did this standard provide the basis for the definition of nominal current- and time-current characteristics, but they made it also possible to define the procedure of tests and their evaluations, as well as specific terms. The drawing of the company Wickmann provides a glance at the DIN-standards on the construction types that applied in 1943:[32]

Pos	d	l	Bezeichnung	Bemerkungen
1	22	51	DIN E 41572 (in Vorbereitung)	
2	8	85	DIN E 41569	
3	7	30	DIN E 41570	
4	5	20	DIN E 41571	
5	20	50	DIN E 41573	
6	5	20/25/30	DIN E 41574	
7	6,5	22	DIN E 41579	
8	5	20	DIN 49398	
9	6/7/6	17/17/25	DIN 72581	
10	4,8	30	DIN 72581	
11	20	50	KM 5377	
12	21	585	Grobfeinsicherung	
13	8	42	Kondensatorsicherg	
14	5	20	"	
15	6	40	"	
16	7	80	"	
17	20	65	Hochspannungssicherung	

Construction types of safety fuses (1943)

---

[32] Archive Company Wickmann 2006, Copy in possession of the author.

The "reasonable" dualism of VDE-regulations and DIN-standards was described by a participant of the VDE-0820 committee´s post-war conference in 1953 as follows:

> „(...) dass im allgemeinen die Normblätter die konstruktiven Angaben enthalten, während in die VDE-Vorschriften die elektrischen Daten und die Prüfbestimmungen aufzunehmen wären."[33]

((...) that the standard sheets mainly contain the technical specifications, whereas the VDE-regulations should be complimented with the electrical data and test conditions)

The regulation for the standardization of radio sets, which was enacted by the chief representative for technical communication media in July 1939 allowed the implementation of a significant reduction of the nominal currents in accordance with the standard. Up from then, the former common and individually adjusted limiting currents and maximum load currents were replaced with the term "nominal current" and reduced from the former average voltage rate 45 down to 15. This brought about significant consequences for the manufacturers of e.g. radio sets. The alternative of a fuse that balances a possibly incorrectly adjusted overcurrent was only possible in the form of a "specially designed model".

4.3. Limiting current or nominal current?

Due to its resistance, every conductor that is flown through with electricity converts electrical power into heat. The conductor (fuse-element) returns the generated heat quantity Q_{zu} by way of heat radiation, convection, and heat conduction to its surroundings and contacts. A significant aspect is the heat conduction through the fuse element to its surrounding medium (e.g. air or sand) and its current supply lines. The heat output Q_{ab} is limited. As long as the discharged quantity of heat remains higher than the generated heat quantity $Q_{ab} > Q_{zu}$,the fuse wire does not melt. If the ratio is reserved, i.e. $Q_{ab} < Q_{zu}$ the fuse element melts within a specific time t and the fuse and shuts off the current supply. Another important aspect to be considered is the steady state, if $Q_{34} = Q_{zu}$. In cases like this, i.e. the highest possible load, the fuse element would be capable of conducting the electricity to infinite lengths. Nowadays this kind of current is referred to as "limiting current". Since the heat conduction to the connecting contacts is dominating, the limiting current is strongly dependent on the dimensions of the contacts and the diameter of the fuse element.

---

[33] Protocol 1. Conference of the VDE-committee 0820 „Gerätesicherungen der Fernmeldetechnik" (Safety fuses in the field of telecommunications) on the 11.2.1953 in Bacharach. The objection filed by the VDE against the standardization of the characteristics of safety fuses stated in the DIN-sheets 57161 and 41574-41477 back in 1952, led to a harmonization of the standards

This correlation was found by W.H. Preece in 1864, when he conducted several tests with cables. The mathematical formula developed by him is referred to as the "Preece´s Gesetz" (Preece´s law):

$$I = a * d^{3/2} \text{ (a= Material constant; d = Diameter)}$$

However this formula applies only to infinitely long lines (l>200 mm). With shorter lines the heat conduction flowing into the contacts of the fuse elements becomes the dominant variable. They shift the limiting current to considerably higher levels. Hence, it can be said that a safety fuse must be aligned with the characteristics of the device in order to provide for maximum protection.

The way of how "nominal current" is defined is explained by means of the following example: A load current exceeding the nominal current by 1.5 times must have a resistance of one hour, whereas a load current exceeding the nominal current by 2.1 must be shut off within 120 seconds (Standard IEC60127-2-III, slow safety fuses 2003). Provided that all of these requirements are met, the nominal current applied for the device is deemed to be approved. In order to provide a basis for the comparison of the test results from different test sites, these standards contain particular test conditions. When it comes to the definition of the nominal current, it is necessary to first determine a sample holder whose mass is large enough that the sample holder can be considered as an "ideal heat sink". This comes along with the conclusion that a heat compensation ($Q_{ab} = Q_{zu}$) will actually never be possible.

The terms „nominal current" and „permanent load current", which were common among the users until the 1930s are rather related to the application or device itself, whereas the term "nominal current" (similar to other characteristics of the safety fuse) relates to a standardized value, so as to provide for both a consistency and comparability. The number of nominal currents set out in the VDE 0820 was first determined at max. five or seven nominal currents per each decade (between 1 and 10), e.g. (1,2 A); 1,6 A; (2A); 2,5 A; 4 A; 6 A and 10 A, where as the nominal currents 1,2 A and 2 A were supposed to be avoided. Even in 1948 the nominal currents stated in the Wickmann catalogs were neither categorized nor did the model follow any systematic approach. It was not before the year 1955 when the Wickmann engineer Hermann Bellen described a gradation of the current steps in accordance with the standard DIN 323[34] of the set of rules R10.[35]

---

[34] DIN 323, Issue 1/1922: Normungszahlen, Millimeter.(Standardization codes, millimeters)

The standard figures or Renard-Series are preferred numbers based on geometric sequences and specified in the standards ISO 3 and DIN 323. In mathematical terms, these are defined by means of lines through the multiplier: $Rm = \cdot \sqrt[m]{10}$. An m in the form of m = $3*2^n$ (with n ε {0, 1, 2, 3, ...}, thus m = 3, 6, 12, 24, ...) delivers values, which are based on the E-lines for electrical component parts. The most commonly used values for m are 5, 10, 20 or 40. For simplification reason in practice, the standard values are usually rounded off.

It seems that the realization of an ideal, individual device protection (as it was common practice until the enactment of the standard VDE 0820 in 1943) under consideration of all standards has become impossible, unless those standards are not kept and practiced.

The post-war years were characterized by the striving to break into international markets through standardization and in so doing pave the way for the serial production.

---

[35] Bellen, Hermann: „Die Gerätesicherung" (The safety fuse), Witten 1955.

5. 1946 to 1970: Standardization and serial production

A safety fuse without a standardized design and characteristics, nor certifications, the necessary serial production and associated breakthrough in the European and international markets would not have been possible. After all, the at least partly automatic serial production of 30 different fuses with individually aligned "limiting currents" in about 20 different construction types turned out to be quite difficult. Since safety fuses were supposed to be built and applied all around Europe or even worldwide, a standardization of them was indispensable.

In England, not only the IEE (Institute of Electrical and Electronics Engineers, 1872), the IEC (International Electrotechnical Commission, 1906) but also the BSI (British Engineering Standards Association, 1901) tried to establish their systematic approaches regarding a standardization since the end of the 19th century. In the USA, the UL ("Underwriters' Electrical Bureau", 1894) standardized the engineering methods and test procedures. The same goes for Switzerland by way of the VSM-standard („Verband Schweizerischer Maschinen-Industrieller", 1939). In 1893 the VDE (Verband deutscher Elektrotechniker) (Association of German Electrical Engineers) was founded in Germany and the first DIN-standard (German Institute for Standards) came into effect in 1920.

Yet, it was a long way until the implementation of a systematic standardization of safety fuses. As described above, a definition of the formerly common laminated fuses used for power outlets in 1925/1926 can be found in the standard sheet DIN VDE 9398 (as of 1941 DIN VDE 49398), which was merely a small step towards the establishment of a "standard system" given the constantly changing demands on the design (e.g. with respect to the dimensioning).

Since the 1930s, the technicians and engineers of the company Wickmann, which is the market leader in the field of safety fuses have had a major impact on the definition of a "standard system" as such. This goes especially for Oskar Ackermann who was employed with the company Wickmann until 1940 (up from 1940 with the company Pudenz) made a great contribution to the preparation of the VDE-regulation from 1943.

In August 1952, the chief engineer Hermann Bellen (company Wickmann) introduced his proposal for the VDE 0820, which replaced many parts of the previous DIN-sheets. After the VDE got the permission to take up their work again in 1952/1953, the proposal by Bellen served as a template for a revision of the VDE 0820, which dates back to 1943. The extent of how far the template was adopted hasn't been figured out, yet.

In 1955 the IEC, which was founded back in 1906, established work groups for their sub-committee SC 32C in order to enhance the work and completion of tasks of the commission. Germany was called upon to join the cooperation. One of the assistants was the graduate physicist Christian Gutzmer, who worked as the head of the research and development department at Wickmann. In February 1974, he was appointed as the convenor of the SC 32C.

One of the work groups of that sub-committee, namely the Working Group WG3 concentrated on working out a definition of "safety fuses for printed switches". Owing to Gutzmer, Wickmann made a significant contribution to the fundamental work and coordination.

Actually, the circuit board or "printed switch" was invented by Charles Ducas in America in 1925, yet it wasn't until 1942 that it gained importance, when Dr. Paul Eisler filed a patent application for the production of a circuit board and described it as a "Kupferfolie kaschiertes, plattenförmiges Isoliermaterial" (plate-type insulator with a copper foil lamination) (pre-patent 1936).

The use of the circuit board gained in popularity after the invention of the transistor. Although the semiconductor technology was discovered by the German physicist Walter Schottky (1886-1976) already in 1939, the use of the vacuum tube was the preferred option until 1957. It wasn't until the middle of the 1950s until the use of transistors became more frequent, whereas the use of semiconductor parts (transistors, diodes) in the form of vacuum tubes became more common in the early 1960s. This provided the essential basis for the miniaturization in the fields of electronics,[36] which also had an effect on the construction design and engineering of safety fuses.

---

[36] The invention of the transistor, http://www.leifiphysik.de <8.2.2012>.

6. 1971 to 2006:
The challenge: Microfuses for "printed" switches

Up from the early 1970s, the new focus in the field of safety fuses was on the miniaturization and alignment of their characteristics with the new requirements of the circuit board technology. Those requirements were pretty high, since the selectivity of the nominal current grading had to be observed, but the use of a standard "limiting current", which was adjusted and tailored to the protection requirements of the respective device was not possible. It was easier to generate a nominal current of 0.032 A or 10 A in a tube with a length of 20 mm, than in a tube with a length of only 5 mm or later by means of SMD-fuses, whose length did not exceed a length of 1.6 mm.

Nevertheless, at least the companies, whose representatives were involved in the WG3 project were striving to develop safety fuses for the circuit board. Some of them used to provide the common "tube fuses" (e.g. 2x20 mm safety fuses) with connection wires so that it was possible to mount them onto a circuit board. Wickmann was rather looking for an innovative, new construction type.

The result was the development of a "Kastensicherung" (box fuse) abbreviated as K10.

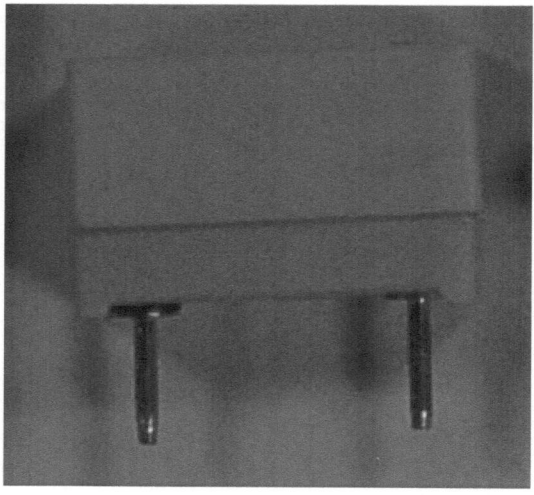

Box fuse K10, Image: M. Rupalla

The distance of the connection pins was at 10 mm (the common term in the circuit board technology, which was used to refer to distances was "4e"). Basically an automatic processing of the safety fuse would have been possible, but the fuse was never intended to be used in mass operation. As the requirements of the standard for safety fuses were adopted for the most part (such as a voltage of 250 V AC, a switch capacity of 1500 A AC and others), this fuse was considered as a high-tech product and was thus, very expensive. Also the ceramic housing and the elaborate design of the contact pins were way too complex and expensive for a serial production. In 1980, the new CEO Dr. Wolf-Dieter Oels proposed a new concept. Owing to the efforts of the SC 32, it was possible to realize the return to a construction type, which was based on a plastic housing. In general, the demands on the performance with respect to a "circuit board fuse" were never as high as those that were put on safety fuses – e.g. the occurrence of a breaking current with 1500 A AC is rather rare when it comes to circuit boards. The trends concerning the development of circuit board fuses in the 1970s and 1980s took two different directions: One of them was focused on the development of a fuse with a plastic housing and a smaller distance between the connection pins (5 mm = 2 e). To the best of my remembrance and based on the research regarding this topic, the CEO Dr. Oels proposed to call that construction type "TR5", which was based on the housing shape of transistor TO, which was used back in those days. Since Wickmann was, due to Gutzmer, the head of development, intensively involved in the SC 32C (tests, measurement series, and construction drafts were primarily prepared at Wickmann), the standard sheet Part 3 of the IEC 127, which was revised in 1986 enabled the standardization of the performance specs of the miniature fuses (DE 000008411568U1) to such an extent as it was provided by the specs of the miniature fuse, which was developed and patented by Wickmann in May 1981 (publication DE 3118943) resp. registered as a petty patent in April 1984 and in accordance with the results of the test series.

Coincidence? Far from it! A correlation is reflected in the mere fact that the TR5 had no competition for years, which is due to the standardization and extensive patent protection. Yet, it must be mentioned that the success of the TR5 was also facilitated by the side benefit that it paved the way for an innovation in the fuse element technology, i.e. the development of the TR5 as a slow model type. After all, the micro- and picofuses, which were developed by Wickmann on the basis of a little fuse license back then had (extremely) rapid switch-off characteristics and were, apart from that, only compatible with the U.S. standard nominal voltage of 125 V AC.

The patent by the company Olvis, which was based in the Netherlands, turned out to be very helpful for the development. Back in 1973/1974 G.I Deelman, who worked with the Olvis had applied for a patent for fuse elements, whose wire was wrapped up around an insulating core (US3845439) and provided with balls of the reactant (usually tin), which were required to achieve the M-effect (US385814). In the course of their own researches and development works, Wickmann replaced the balls with a continuous, galvanically mounted tin layer, in order to circumvent the patent.

Over the course of the following years, the TR5 was further improved including variations of the construction design for non-standardized applications (TE-protector), and the protection of the manufacturing method by means of additional patents. Those patents were especially related to the production engineering and that way prevented a replication, which would have cemented the success of the TR5. There was one thing that couldn't be realized by hook or crook: It was impossible to apply the TR5 along with the SMT equipment for a circuit board, which was introduced in the 1980s. Such SMT circuit boards were only compatible with SMD component parts. These parts weren't soldered and plugged into the circuit board by means of connectors, but soldered directly onto the circuit board. The following figure shows a circuit board, which is equipped with SMD component parts.

(Image: M. Rupalla/Wickmann-Laboratory) SMT-circuit board[37]

With the afore-mentioned second development facility, Wickmann made it to prove their sense of foresight and made provisions. The goal of one project in the course of this development series was the production of printable fuse elements in the thick-film technology.

---

[37] Image: http://www.leiterplattenbestückung.net <8.2.2012>

Already in the 1970s, Wickmann conducted a development project that focused on the investigation of possibilities with respect to fuse elements used in the field of thick-film applications. The goal was however not the miniaturization of the fuses for the circuit board, but the replacement of the ultra-thin circuit board wires for fuses that had lower current grade levels (In < 100 mA with D < 0,01 mm). When used for the generation of thin layers, the evaporation method turned out to be rather complicated and very expensive. Besides, it was nearly impossible to process the very thin layers (D < 0,005 mm, which led to the discontinuation of the project quite soon.

The evaluation period of the thick-film technology, which aimed at the production of layer fuse elements took place at almost the same time as the development of the TR5, which was back in 1981. The thick-film technology is based on the methods applied for the silk screening and was used to provide – mostly ceramic – brackets with electrically conductive materials or insulation materials. As it was also possible to apply reactants (solder paste), the use of a "slow" fuse element was basically no problem, although the inertia required in the standard sheet 127-3 wasn't achieved. Wickmann had this construction design for fuse elements patented in 1982 (publication DE 3044711 A1). Even today, the basic principle of the Ag-Cu-Sn-Reaction protected in the patent claim is still in use for slow chip fuses.

The development project "Layer safety fuses" involved several products and patents. In response to the SMD component parts which had been in use since 1982, Wickmann introduced the worldwide first chip safety fuse, namely the Wickmann Thick-film fuse, whose construction design was entirely based on the SMD-technology at the exhibition "Electronica" in 1986.

Chip safety fuses; Image: M. Rupalla

The chip with the dimensions of 1210 (3.2 x 2.5 mm) had a rapid or extremely rapid fuse element using the silk screening technology.

In the following, this method and its peculiarities will be described in detail. The success in the past and nowadays left apart; whereas the TR5 is "merely" the next generation of the technology using wired fuse elements, which has been in use since 1880, it is a matter of fact that the layer method paved the way for entirely new construction designs, protection concepts, and above all, an innovative individualization of the overcurrent protection without the abandonment of a serial production.

The method to develop layer fuses applied at Wickmann was characterized by its two or three layer fuses system. Leaving the first phase of the above-described thin layer fuse elements unconsidered, it can be said that the foundation of the layer fuse elements or layer fuses was laid between 1981 and 1987. The concepts revised and improved from 1995 to 2002 were more elaborate and aligned with the market situations.

7. The basic stages 1981 to 1987

Back in the 1970s the goal of the „thin layer projects" was the replacement of thin fuse element wires, but now the aims of the project were rather focused on the establishment of new technologies with respect to an innovative protection concept. The emerging SMD-technology and the potential miniaturization of the electrical systems gave the decisive impetus for new approaches with respect to the safety fuse technology were the

The silk screening technology, which was already used in the manufacturing processes of the first "hybrid switches" (silver conductor tracks printed onto ceramic plates with printed resistances and soldered SMD-semiconductors) was very advanced and enabled the realization of very precise conductor tracks (conductor track width up to 0.1 mm width and 0,003 mm thickness). Already back in 1980, the definition of future goals took place in the course of a project that was funded by the Federal Ministry for Research and Technology (BMFT):

Development of thick-layer fuse elements, which are suitable for serial production using the silk screening technology.

1. Individual alignment of the fuse element's nominal current or limiting current using a resistance balancing by means of a laser or sand blast technology.

2. Development of chip fuses using a metallization of the chip edges for the surface mounting (SMD-chip).

3. Development of layer element fuses with slow shut-off characteristics (Basic principles for the works related to the standardization carried out by the SC 32C).

4. Integration into and interaction of the fuse elements with the switching circuits to be protected.

The final report was submitted to the Federal Ministry in the middle of the year 1984. Most of the set goals were achieved under consideration of the following approaches:

On 1. Depending on the defined chip dimensions, it was possible to print more than 1.000 fuse elements within one printing process. Two or three layering processes were necessary though, but since it didn't take longer than a couple of seconds to complete one layering or printing process and the respective process was generally automatable by way of simple, customary means, it was possible to realize a serial production that complied with the requirements and standards.

On 2. As already described above, there's each one defined resistance that can be assigned to the nominal or limiting current of a safety fuse. By aligning the resistances, the high precision of the silk screening technology could be increased to a considerable extent (aligning = targeted, locally restricted lancing of the printed fuse elements using a laser or by means of reducing the fuse element's thickness by means of a sand blast).

This work step was for the most part integrated into the automated production line.

That way, it was possible to draw back on the formerly common individual alignment of the fuse with the respective residual current to be shut off, without having to spare an automated serial production.

On 3. The metallization of the chip edges was already tested and approved in the course of the serial production of chip resistors and chip condensers and could be implemented almost fully automatic.

On 4. The manufacturing of thick-film conductors with slow shut-off characteristics (slow time-current-characteristics), which were manufactured in accordance with the state-of-art of the standard IEC 127 Part 3 using the heat-conductive ceramic plates (substrates) couldn´t be implemented. Apart from that, the realization of a shut-off delay in case of slight over-currents was possible (see above).

On 5. The criterion represents the actual core of a new philosophy in terms of safety fuses. Both the thick-layer and the silk screen fuse elements were applied on a heat-conductive ceramic plate. The heat conduction and the precision of the silk screening technology suggested the installation of a resistor in close proximity to the fuse element.

Provided that the material chosen for this resistor was subject to deviations with respect to its resistance capacity when temperature fluctuations occurred (NTC-resistor material), the current load could be derived from the respective resistance value. In cases where the selected resistance threshold was below the desired value, an electrical system could be applied in order to take the measures required for the handling of that load (e.g. shutting off the current). In 1983, the developed component part was patented as "Protensor" (word combination of "protector" and "sensor") (DE 3221919 C2). The actual fuse element could be construed as an optimized short-circuit protection.

However, there was found a material that was resistant to temperature fluctuations. That way, it was made possible that the resistance in a defined current flow generated a temperature that led to the melting of the tin-coated fuse element. The component part used for this purpose, i.e. the triggerable safety fuse was also patented in 1933 (Austria 383 697 B).

The realization of the interaction between the safety fuse and the electronic device to be protected provided the possibility to integrate the safety use into intelligent protection concepts, yet, it wasn´t made use of. Later on in 1987, the company Wickmann discontinued the work on the project "thick-layer safety fuse".

In 1995 the project was revived on the basis of the principles established in the 1980s. The hybrid switches designed for the purpose of delivering high performance along with a high precision at high temperatures (especially in the field of IT- and automotive technology) brought about an improvement in the thin film technology and high precision. Innovative bracket materials (substrates) and new silk screen-compatible insulation material would have made a restage of the former projects appear to be reasonable.

Although Wickmann had left their competitors in the field the development of chip fuses behind long ago, and the competing products didn´t get beyond the simple SMD chip fuses, the idea of an integration (the interaction of safety fuse and electronics (see e.g. "Protensor" or "triggerable safety fuse") which were popular in the 1980s regained popularity. Besides, the small dimensions made it even more difficult to control the SMD fuses in their common wiring technology. Above all, an immediate integration of the wiring technology into an electronic system was hardly conceivable.

Therefore, the realization of slow SMD safety fuses in accordance with the standard sheet IEC 127 Part 4 with diminishing size regarding the wiring technology was not possible. After the revival of the thick-layer projects up from 1995 until 1997 which was based on the principles from the 1980s, it didn´t take long until the patenting of the worldwide first slow chip safety fuse (DE 197 04 097 A1) that complied with the standards of IEC 127 Part 4. This was regarded as a little sensation from the viewpoint of the experts in this field.

Apart from that the miniaturization of the chip fuses was optimized and its combination of thick- and thin layers brought about the realization of chip dimensions of up to 0603 (1.52 x 0.76 mm).

However, the attempt to boost the serial production of the new chip fuses in a most cost-effective way failed. The same goes for the customer-specifically designed safety fuses in accordance with the standards of the hybrid technology, as the timing was simply too late. Many users fell back on other solutions. Due to this reason, the project "thick-film method" was abandoned in 2002. It was the last ambitious effort with respect to developments in the field of fuse engineering, which would have completed the extensive and advanced product line by the company Wickmann.

8. New approaches in the research until 2012

The attempts of an integration into the circuit board, the "embedded Fuse". Publication on a project, which was funded by the DBU (Deutsche Bundesstiftung Umwelt) in the technical magazine "Elektronik Praxis" 3/2011.

(Image: M. Rupalla/Wickmann-Laboratory)

8.1. "The conductor track as a safety fuse: Dangerous bricolage or a matter of necessary, individual safety?"

Since the invention of circuit boards, it happens time and again that sections of conductor tracks are construed as a "safety fuse". Today this variant of the over-current protection is frequently applied in connection with a number of applications, which is most often not recognized by the inspecting authorities. It is unknown to which extent these "conductor track fuses" are the cause for fire damages on devices. There is not much known about the reasons for their use, either. The answer on the question whether this is due to the insufficient knowledge on the part of the developers, or the high costs incurred with an appropriate safety fuse, or even the lacking adaptability of standard fuses to the application that is supposed to be protected is definitely subject to the individual case..

The performance specs of appropriate fuses are listed in a specification chart, which is among others based on standards. The experiences gained from numerous cases of practicing consultancy have shown that there is a high demand for individual fuse solutions.

The development of a customized fuses usually involves great effort, high costs, and is – to put it mildly – not really interesting.

The individualization of the protection functions in a customized over-current fuse can also be achieved by means of a conductor track fuse. Yet, it may be useful to know that the design standards for the development of device fuses apply as well for circuit boards and, that the freedom of design with respect to the desired shut-off characteristics is, indeed, transferrable.

A project funded by the DBU (Deutsche Bundesstiftung Umwelt) and implemented by the company M&M-Elektronik made clear that the design standards with respect to conductor track fuses are generally the same like those applicable for the development of SMD fuses, especially chip fuses.

These design standards can be broken down into three essential requirements.

8.1.1. Alignment of the melting heat onto a locally restricted "hot spot"

The area surrounding the conductor track of the fuse, which starts to melt in case of an error, must be locally restricted in order to prevent a thermal overload of both the basic material and the surrounding areas (Risk of fire!). The conductor track fuse must have a considerably higher resistance than the feeding conductor tracks. Since there is a functional correlation between the hot spot and the desired shut-off/load current, the geometrics and the resistance values of the feeding lines and hot spots must be aligned with each other.

A resistance balancing made this possible. This was realized by means of a Trim-laser or a selective etching technique, which was especially developed for the project.

Images (starting from the left): Etched Hot spot; by means of a laser cut, ore hot spot; shut off hot spot   (Image: M. Rupalla/Wickmann-Laboratory)

Reduction of the melting temperature by means of diffusion processes

The melting temperature of Copper is 1083°C. In case of errors this temperature must be generated in the hot spot, even if it is only for a short time (usually t<1s).

In order to reduce the melting temperature, the hot spot is partly covered with tin. The diffusion or dissolution process of the copper becomes strongly accelerated after the melting point of tin, which is at 232°C, is being reached. Hence, the switch-off temperature is usually limited to T<300°C. The dissolution process takes more time (until approx. t<10s), which brings about a certain sluggishness with respect to the responsiveness of the fuse. In fact, the "slow" or "sluggish" It characteristic resulting from this is usually desired.

Regarding this, it must be mentioned that, depending on the hot spot geometrics, the volume of the tin is attached to great importance. The tin quantity must be sufficient so as to dissolute the copper of the hot spot and the tin must be positioned at the hot spot in such a way, that it´s no problem to use the produced heat for the dissolution of the copper conductor track.

Image: M. Rupalla

Suppression resp. control of potential shut-down light arcs

The tests conducted at M&M-Elektronik showed no mentionable shut-down light arcs at shut-off voltages of 12 V DC, so that it wasn't necessary to take any additional measures to suppress the generation of light arcs. Provided that the hot spot was covered with a silicone of the company Sinus, the results turned out to be well with voltages around $U_A>15V$ DC. However, it is not possible to use a simple cover, as melted material of the hot spot generated relatively low-ohmig conductive tracks under the cover. The performance, which was then achieved over a longer period of time led to a considerable heat build-up of the circuit board underneath the hot spot and a charring of the circuit board material (in this case – standard FR4). In the majority of tests, this escalating process of power dissipation and charring ended up in a fire of the circuit board.

An effective way to control those light arcs was found in the application of a double layer cover, which was often used in connection with chip fuses. The hot spot was covered with a low-melting material. It must be seen to it that the quantity of the material is sufficient to take melted hot spot material.

Good results were achieved with the use of hot wax, which is for example used to fuse component parts. This material serves as a basic layer, before a silicon grouting is poured over it, so that the first layer won't dissolve in case of temporary loads (e.g. during the soldering process).

Economic and ecological advantages of conductor track fuses

Of course there's nothing to say against the commitment imposed on the manufacturers of devices or switches to get involved in new and unknown fields of the electrical engineering: Namely, the over-current protection!

Apart from the fact that the implementation of a really safe over-current protection makes this inevitable anyway, even if no conductor track fuse is used, the learning efforts are definitely worth it:

- More safety thanks to the alignment of the protection
- Less soldering joints and hence less flaws
- Less efforts when it comes to the warehousing and logistics during processes
- Less pollutants with respect to the packing, soldering, and cleaning less efforts with respect to the mounting and assembly

These were merely the obvious advantages.

It is rather unlikely that this is only possible without the approval of the inspecting authorities. A research study by M&M-Elektronik on the acceptance of conductor track fuses based on the design described above, turned out positive to such an extent that the inspecting authorities even hand out test certificates. Yet, licensures or approvals are not possible due to the (yet) missing part characteristics, e.g. dimensions, solderability, etc.

All electrical specifications were achieved in a test according to the standard IEC 127-4.

Depending on the type, it is possible to achieve a shut-down capacity of 24V/35A DC. The possible It- characteristics are depicted in the figure below:

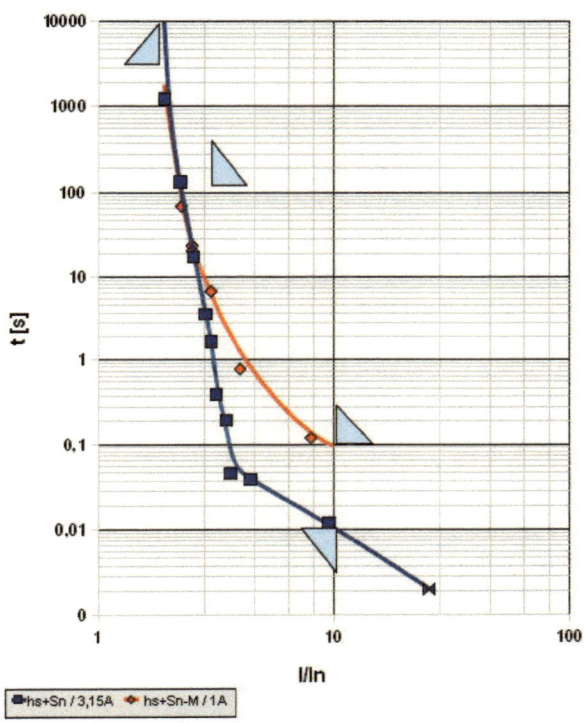

DBU-Projekt/Kennlinienvergleich verschiedener Leiterbahn-Schmelzleiter IEC 127-4 (träge)

Conclusion:

The company M&M-Elektronik puts many self-developed applications to the test so as to find out whether they can be used in connection with a conductor track fuse and if a realization of the same would be reasonable. The solution for a conductor track fuse is individually adapted to the requirements of an application. Although many things can be predicted by experience, the conduction of additional tests can´t be ruled out, and those may turn out very differently, as the factors requirements and experience play a significant role. As it is the case with the selection of standard fuses, the potential loads and fault switch-offs should always be tested at max. operation voltage, whereas the test scope depends on the respective requirements and previous experience.

8.2. New wire materials (Company Elschukom)

There <u>are</u> only a few possibilities to achieve a slow shut-off behavior in a safety fuse.

- o Constructional, as is known from history, by coiling the fuse element using a combination of the latter, i.e. inductors or resistors or by means of the combination of various fuse element materials and the use of the M-effect outlined above.
- o The combination of various construction types within one fuse.
- o Combination of the fuse element with electro-mechanical or electronic component parts. Some of the most promising types were reflected in the approaches of the layer method (e.g.. Protensor, P.61).
- o Use of fuse element materials, whose specific characteristics enable or at least facilitate a slow shut-off behavior.

Also the latter option is not new and has been put to several tests over the last decades.

The problem was the lack of knowledge about the material characteristics and their interaction, when the used fuse elements were put under load (heat built-up).

Until 1945, the research was rather based on empirical findings, whereas the development was rather oriented towards the application. The standards that gained importance after 1945 brought about a prioritization of the standardization of partly not fuse-specific characteristics (shape, dimensions, test conditions and requirements, ...). The basic research, to that extent as it was open to the public, was based on empirical findings, hardly systematized, and rather focused on mere approaches.

Up to this day, the significance and meaning of the material characteristics are rather rooted in the experience of the developers than in documented knowledge.

Of course, there are countless diploma theses and dissertations that were published after 1945, which dealt with the various impacts of material characteristics (above all the M-effect). Yet, this information is not part of the teaching material at colleges and universities.

What has been found out, is that the varying diffusion behavior of different metals may not only generate but also prevent sluggishness.

It is known that the combination of different metals – whether in the form of an alloy or in the form of a compounded fuse element - may react in both ways that is either rather slow or rapid.

When you ask a "developer" of fuse manufacturers, he will most probably come up with an explanation for those effects. It goes without saying that it doesn't make much sense to ask several developers, as the viewpoints vary and so do the explanations (empirical knowledge).

However, this doesn't turn out to be the optimal basis for the research or development of materials with ideal and controllable characteristics.

The company Elschukom has attended this deficit and tried to find profound and universally valid explanations by means of selective basic research projects.

The recent status quo until the end of 2012 is described in the following. As some of the projects haven't been finished yet, the descriptions are limited to that extent as is required to provide a general overview.

8.3. New theoretic approaches (Company Elschukom)

Analyses of the thermal conditions in safety fuses in switch-on mode showed relatively complex correlations between the generation, radiation, and accumulation of heat. This interaction or the mutual conditionality is often referred to as thermal balance. When it comes to the respective bracket (e.g. standardized sample holder) the thermal radiation to the housing, contacts (left aside the usually lower heat generation due to the contacting), and surroundings (e.g. filling media) of a production series remains mostly at a constant level. Depending on the fuse type and nominal current, the heat generation through the fuse element may vary considerably.

Basically the determining characteristics of the function of a fuse are reflected in the attributes geometry and material of the fuse element. All essential (partly standardized) characteristics of the fuses, such as the resistance, nominal current, and shut-down characteristic are defined by the fuse element. Therefore and for the purpose of achieving the desired and required safety of our electrical devices, it suggests itself that it would be quite reasonable to deepen our knowledge and understanding about the functions of material characteristics by means of basic researches. This is the only way to analyze and compare the effects of changes in the materials

while at the same time allowing for targeted research on new, safe, and ideal materials.

A feasibility study conducted by the company Elschukom in collaboration with the Beuth-University Berlin has shown that it is possible to realize a mathematical modelling, which is based on the specific material characteristics, which in turn enables the detection or proof of the influence of various parameters in a qualitative and quantitative manner. The figure below shows an It-characteristic and some of the parameters including their most significant spheres of influence.

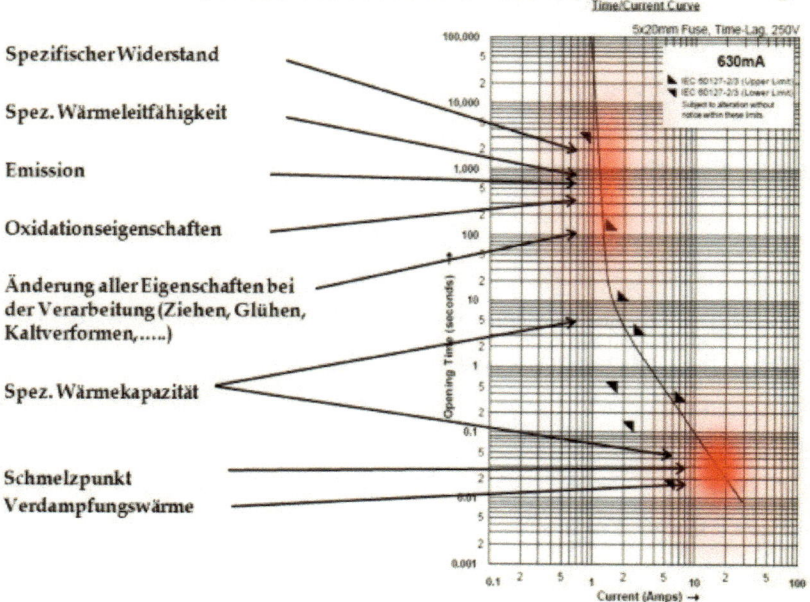

The temperature dependency of several parameters was also considered in the modelling.

The reliability of the mathematical model was proven by means of various materials, which were compared to empirically determined characteristic curves. The figure below shows a comparison of the material copper.

Cu99,9% - Ø 0,150mm

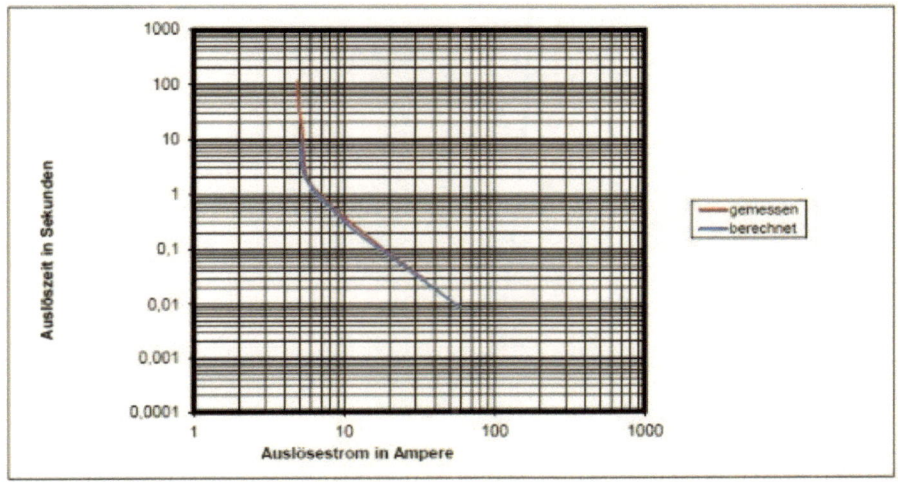

The shut-down sluggishness of a fuse element is often achieved through a tinning of the basic material (M-effect), despite it has been a common fact that the mathematical acquisition of the data concerning the diffusion behaviour of both materials is not possible at all or only with difficulty.

The feasibility study proved that it was possible to integrate the M-effect into the calculations, as is depicted in the following figure with the characteristics of an AgCu-alloy.

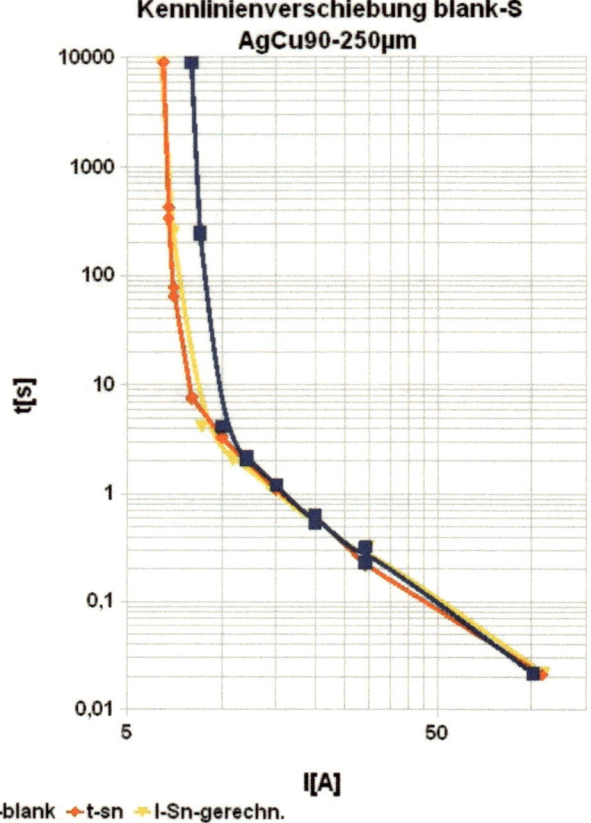

The results of the feasibility study lead to the conclusion, that the realization of an extensive project allows the definition of the characteristics of safety fuses in different materials. That way, it is not only possible to draw comparisons between the fuse element materials, but it also provides for the likelihood of the predictability concerning the impacts by the geometrics and/or changes in the material.

It remains to be seen, whether the calculation and specification of feasible "materials or material combinations on request" will be possible in the near future. The project was planned and initiated by Elschukom in 2013 bei Elschukom. The project is complemented by a bachelor thesis from the Beuth-University Berlin. The research work carried out by the company the involved universities hasn´t been completed, yet, thus it will be interesting to see, what findings will be revealed in the near future.

9. List of references and literature

Ackermann, Oskar: Warum eine Gerätesicherung? Witten 1936.

Ackermann, Oskar: Die Absicherung von elektrischen Geräten, Witten 1934.

Ackermann, Oskar: Die Absicherung elektrischer Anlagen in Luftfahrtzeugen, Witten 1936.

Andrews, Leonard: Electricity Control. A Treatise on Electronic Switchgear and Systems of Electric Transmission, o.O. 1904.

Bellen, Hermann: „Die Gerätesicherung", Witten 1955.

Cames, Wilhelm: Die Normung in der Funktechnik, in: Funkschau 17(1944)3/4, S. 17-18.

Cockburn, Arthur C.: On Safety Fuses for Electric Light Circuits, and on the Behavior of the Various Metals Usually Employed in Their Construction, in: Journal of the Society of Telegraph Engineers 16(1887)5, S. 650-665.

Das erste Radio 1924. Eine Veröffentlichung aus der Freien Presse vom 3.3.1992 http://www.krumhermersdorf.de/geschichte/k-g1631.htm <8.2.2012>.

Degen, Jost, Archiv der Firma Schurter, 2008. Die Erfindung des Transistors, http://www.leifiphysik.de <8.2.2012>. Funkschau Heft 18 – April 1939.

Gelet, Jean-Louis: To the Origins of Fuses, 8th International Conference on Electric Fuses and Their Application, Clermont-Ferrand 2007, S. 1-8.

Federal capital Düsseldorf (Hg.): Risk House fire – Recognizing hazards, Preventing hazards, Düsseldorf 2006 http://www.duesseldorf.de/feuerwehr/pdf/alle/risiko_-wohnungsbrand.pdf<8.2.2012> Meyer, Georg Isidor: Zur Theorie der Abschmelzsicherung, Munich/Berlin 1906.

Protocol of the 1st conference held by the VDE-committee 0820 „Gerätesicherungen der Fernmeldetechnik" on the 11.2.1953 in Bacharach.

Cover sheet of the VDE 0820 from 1943, reprint from 1952, rendered for the registered limited edition by courtesy 202.013 of the VDE Verband der Elektrotechnik (Association of electrical engineers) Elektronik Informationstechnik e.V.. The most significant criterion for the application of the standards are their versions along with the latest edition, which are available at VDE VERLAG GMBH, Bismarckstr. 33, 10625 Berlin, www.vde-verlag.de.

Stiftung gemeinsames Rücknahmesystem: Die Welt der Batterien. Funktion, Systeme, Entsorgung, Hamburg o.J.

http://www.grs-batterien.de/fileadmin/user_upload/Download/Wissenswertes/Infomaterial_2010/GRS_welt_der_batterien.pdf <8.2.2012>).

Telefunken-Zeitung Nr. 17/1919, P. 10

http://www.Radiomuseum.org <8.2.2012>). US-Patent Nr. 622,511 vom April 1899.

Wickmann-Werke AG: Catalog, Witten 1980. Wickmann-Werke AG: Catalog, Witten 1932. Wickmann information on safety fuses Witten 1994

10. Chronological overview on the history of the safety fuse

Company / Name / Event	Year	Time and achievement
Edward Narine	1774	First tests with fuse wires in connection with lightning protection in the field of electrostatics
Louis-Francois-Clement Breguet	1846	Report on the great fire, which destroyed Saint-Germain
Sir William Henry Preece	1864	Telegraphic analysis on the permanent load „protect submarine cables"
IEE „Institute of Electrical and Electronics Engineers"	1884	Establishment of the institute for standardization: Institute of Electrical and Electronics Engineers
Sir David Salomons	1874	First report on the use of fuses
Firma Brush	1879	Brush Electric Light Corporation, yet, safety fuses were not added to the product range before 1957
Joseph Swan (1828-1914)	1880	Light bulb manufacturer, uses tin-foil fuses for light bulbs
Thomas Alva Edison (1847-1931)	1880	First patent for fuse elements used in glass tube housings
A.C. Cockburn	ca. 1880	First thesis on the theory of safety fuses, thermal balance, analogy of thermal-/electrical principles, Switching off between $1.5\text{-}2*I_n(I_g)$
Electrical Exhibition	1881	Electricity exhibition in Paris, large area taken by safety fuses
Extensive test series with "innovative" illumination means	1881	Electrical illumination in Berlin´s downtown
T.A. Edison	1881	First mentioning of the term "safety guard" in a patent
„Electric Lighting Act"	1882	The British parliament adopted regulations concerning the manufacturing of electr. supply units
Oskar Miller	1882	Electricity exhibition in Munich
Charles Vernon Boys (1855-1914), H.H. Cunyngham	1883	Draft and patent of a safety fuse for a solder/coil design
A.C. Cockburn	1887	First written publication on safety fuses addressed to the „Society of Telegraph Engineers"
Firma Brush	1890	Launching of the first illumination unit, without safety fuses back then

W.M. Mordy	1890	First patent for "cartridge fuse" incl. filling and copper wiret
Oskar Miller	1891	Electricity exhibition in Frankfurt
Feldmann	1892	Compilation of initial theoretical basics in the magazine: "Elektrotechnischen Zeitschrift"
VDE „Verband der Elektrotechnik"	1893	Founding of the VDE
UL „Underwriters' Electrical Bureau"	1894	Henry Merrill establishes the „Underwriters' Electrical Bureau"
VDE „Verband der Elektrotechnik"	1895	Draft of a regulation for low-voltage plants up to 250 V (Eisenacher Conference)
BSI „British Standards Institution"	1901	Establishment of the institute for standardization "British Engineering Standards Association" by Sir
IEC „International Electrotechnical Commission"	1906	Founding of the IEC. Among others, the IEC made a significant contribution to the standardization of dimensions
G.I. Meyer	1906	„Zur Theorie der Abschmelzsicherungen", first comprehensive basic guide on fuses published in Germany. Only over-current
Firma Pudenz	1909	Wilhelm Pudenz GmbH is founded in Wuppertal
IEE „Institute of Electrical and Electronics Engineers"	1910	„IEEE's roots, however, go back to 1884". H.W.Kefford: Definition of I_n, U_n, Characteristics of switch capacities and test conditions
Firma Bussman	1914	The company Bussman starts to provide repair works on safety fuses
DIN „Deutsches Institut für Normung"	1918	First DIN in the form of DIN 1 (Cone pin for MGs) and put into effect in 1920
Firma Wickmann	1918	Until approx. 1928 repair of safety fuses
First radio station	1920	Launch of the First German radio station "Königs Wusterhausen"
IEE	1924	Establishment of the institute for standardization „Regulation for equipment of buildings"
Charles Ducas, circuit board	1925	Application for a patent filed at the U.S. Patent Office, used for the first time in 1942
Firma Littelfuse USA	1927	Since the establishment of the company by Edward V. Sundt in 1927, Littelfuse stands for innovation.
„Elektrotechnische Zeitschrift"	1929	Increasing significance of the use of automatic fuses
Firma AEG	1929	First "Installation-Automatic Switches"

Firma Wickmann	1931	Cartridge-Fuse, Laminated fuses High-precision fuse
Radio exhibition	1933	Introduction of the "Volksempfänger" in Berlin
Firma Schurter	1933	Probably repair of fuses, manufacturing works didn´t start before 1937 n
O. Ackermann	1936	Paper on the necessity of a standardization of safety fuses, until then – application oriented standardization of radio sets
Regulation by the chief representatives for technical communication media	1939	Standardization of radio sets
W. Schottky	1939	Discovery of the semiconductor-effect by W. Schottky
VDE-Blatt 9398	1925	"Standardization" of high-precision fuse 5x20 mm
DIN 49398	1941	Adoption of the VDE-sheet 9398
first VDE-Norm for devices resp. telecommunication fuses	1943	VDE-0820, sets requirements and criteria for tests
W. Cames	1944	Article about the regulation concerning the standardization of radio sets
DIN 41571 ff.	1944	Standard for undistinctive and distinctive 5x20/25 mm fuses – Standard nominal current product line limited to max. 6* nominal current
Technical standardization committee "Electrical engineering"	1946	First conference upon approval of the occupation forces
Company SIBA	1946	Replacement of old fuses (repair and maintenance)
Start of the harmonization efforts	1950	Harmonization of national and international standards
FNE 345.1	1950	The (first?) conference of the institute for standardization on the topic "fuses" in Bacharach
E. Wintergerst	1950	On the melting time of safety fuses, published in the " Zeitschrift für angewandte Physik"
H.W. Baxter	1951	Reference book incl. description of switch capacities
Company Wickmann	1955	Approval to join collaboration with the IEC
Company Püschel	Ca. 1960	The company „Püschel Feinsicherungs KG" was founded by the former Wickmann employee Karl Wilhelm Püschel
Company ESKA	1960	Establishment in the 1960s with special fuses and others

Company Ferraz	1969	In the field of safety fuses rather known as distributors
Company Wickmann	1971	(Thin-)film safety fuse
Company Olvis, Deelman	1971	Coil fuse element
Company Olvis, Deelman	1973	Coil fuse element with Sn-balls
H.F. Borchart	ca. 1979	Wickmann technical magazine
H. Johann	1982	Technical book on fuses, focus NH-safety fuses
Company Wickmann	1982	Since 1980 works on the development of the safety fuse-TR5 as replacement of the "box fuse " (IEC SC32C – WG32)
Company Wickmann	1984	Utility model DE000008411568U1 "Miniatursicherung" (Miniature fuse for 250 V AC "TR5")
New construction design „SMD" up from	1985	Increasing demand for SMD-fuses
Company Wickmann	1986	„Electronica": WTF-Thick-film-fuse chip, launching of the first SMD-fuset
IEC	1988	Publication of standard sheet 127-3 "Sub-miniature fuse-links" as a result of the achievements of the SC32C
Company Elschukom	1990	Wickmann-CEO Dr. Poerschke is appointed as the CEO of Elschukom and changes the company´s focus to the manufacturing of fuse element wires and special fuses.
Company Wickmann	1995	Registration of trademark for TR5 for 2911511
Company Wickmann	2004	US-based group Littelfuse acquires Wickmann
Littelfuse	2007	Wickmann is abandoned as a manufacturer and developer of safety fuses

11. Work places through the ages of time

The 1930s

Manufacturing of device protection fuses

Probably a factory walkabout through the pre-fabrication unit

Manufacturing unit for device protection fuses

The 1950s until the 1970s

Manufacturing unit for device protection fuses decades later

A glance into an engineering office

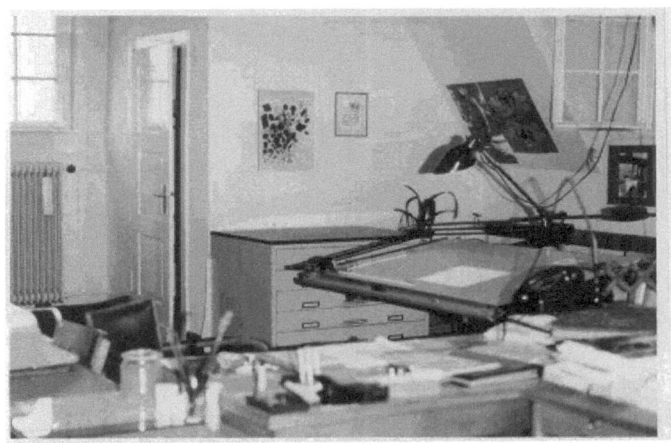

Work place in the "advertising department"

Bild 11 Ansicht der Anlage

Testing plant for the generation of thin metal layers

The 2000s

Manufacturing unit for coiled fuse elements

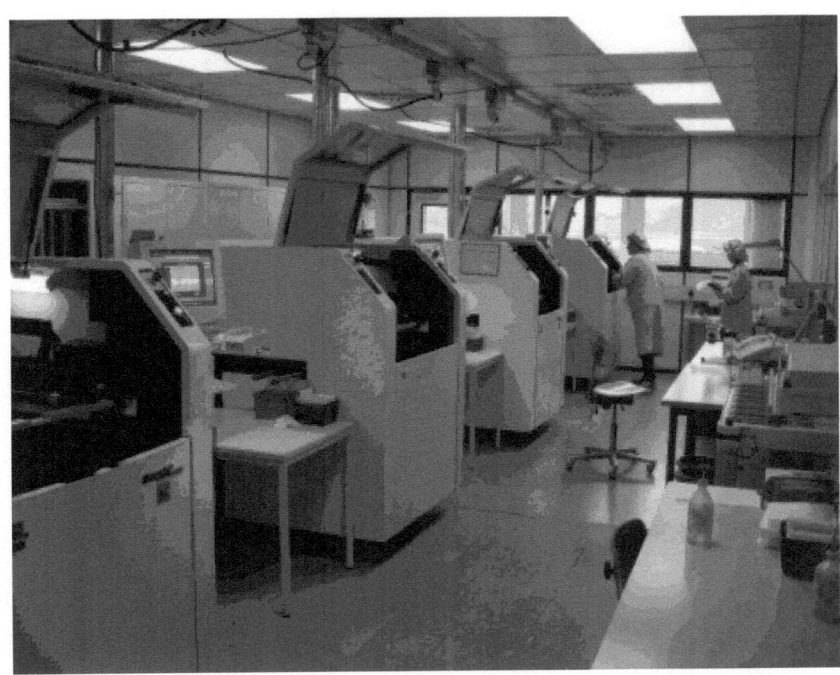

Manufacturing "cleanroom" for fuse elements used in the silk screening technolgy

Work place in the test laboratory of the development department for safety fuses

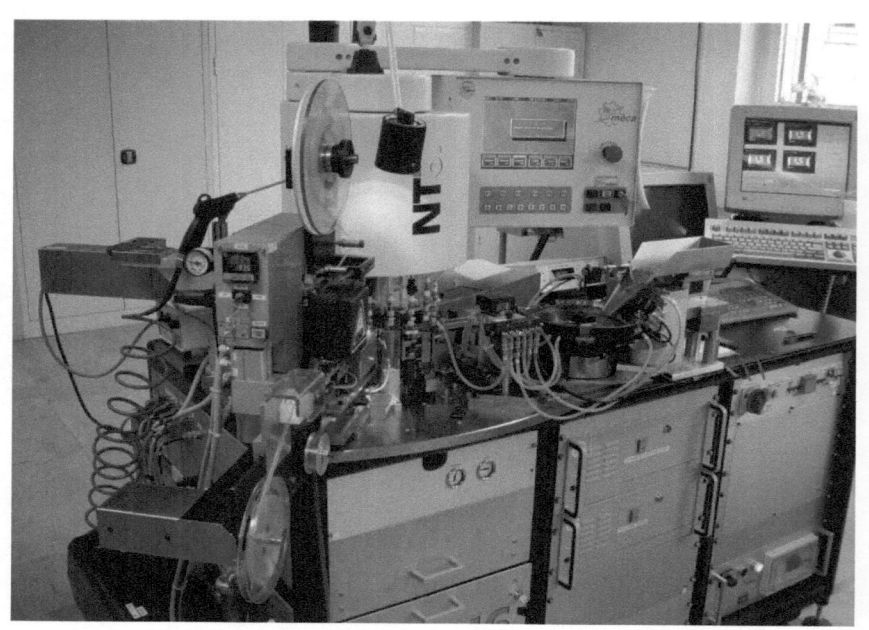

Fully automatic machine for the packing of belts for chip safety fuses.